STARGAZING FOR BEGINNERS

A BINOCULAR TOUR OF THE NIGHT SKY

WITH MAPS AND DIRECTIONS TO HELP IDENTIFY
THE CONSTELLATIONS AND THE BRIGHT STARS
VISIBLE TO THE NAKED EYE

By
The Editors of *One-Minute Astronomer*

Based on the book
"Astronomy With An Opera Glass"
by Garrett P. Serviss

Originally Published By
D. Appleton and Company
New York, 1890

This Edition Published By

Mintaka Publishing Inc.
Copyright © 2009

Introduction To This Edition

This book introduces you to the bright stars and major constellations, along with dozens of deep-sky sights of interest within each constellation, such as galaxies, binary stars, nebulae, and star clusters. It assumes you are equipped with nothing more than a simple pair of binoculars, and that you know nothing of astronomy or the layout of the night sky.

Once you tour the sky through all four seasons with this book, not one person in a thousand will know more about the night sky than you do.

This book is largely based on the classic work of popular astronomy entitled *Astronomy With An Opera Glass* by Garrett P. Serviss, who was the preeminent astronomy popularizer of his day. The book was first published in 1888 to wide acclaim, and it found a place on the shelves of thousands of stargazers and naturalists in the late 19th and early 20th century.

While many books from that era are hopelessly dated in substance and style, *Astronomy With An Opera Glass* stands out because of its accessibility and its casual, friendly approach to touring the reader through the night sky.

Because the stars change their position in the sky slowly compared to the time scales of human affairs, the sky tours in *Astronomy With An Opera Glass* are as accurate and useful today as they were 120 years ago. A stargazer of 1880, if transported to the modern day and armed only with binoculars, would instantly recognize the stars and constellations as essentially unchanged.

But the astonishing advances in astronomy over the past century have badly dated most of the scientific explanation in the original book. This edition fixes that problem. It includes a complete update of the science related to the stars and astronomical sights described in the book. You get completely up-to-date explanations of the science of astronomy, combined with the historical explanation and easy charm of the Garrett Serviss' original work.

And while Mr. Serviss suggested the use of inexpensive opera glasses in his original work, largely because of their low cost and wide availability at the time, you will want to use binoculars. It's easy enough to find a pair, whether you purchase or borrow. Binoculars are superior to opera glasses in magnification and light-gathering ability. And they are far less expensive in real terms than they were in the late 19th century.

Since the maps in the first edition were made with wood cuts, which had artistic appeal but sacrificed accuracy and detail, all twenty maps in the course have been updated using the application SkyX by Software Bisque Incorporated, and Stellarium version 0.10.6. SkyX is a professional-grade application, but Stellarium is free, and it's a wonderful tool for all amateur astronomers. While it's not required to use this book, you can download Stellarium at www.stellarium.org. It's a great star atlas and learning tool and you'll turn to it frequently.

While this book makes it easier for you, it still takes a little effort to learn the stars. Take it slowly, and just try to learn a little more each night. After a few nights, you will be amazed at how much you've learned. And above all, as Mr. Serviss says,

"Do not be afraid to become a star-gazer. The human mind can find no higher exercise. He who studies the stars will discover, 'An endless fountain of immortal drink; Pouring unto us from heaven's brink.'"

Brian F. Ventrudo, Ph.D.
Ottawa, Canada, May 2009

Introduction To The Original Edition

In the pages that follow, the author has endeavored to encourage the study of the heavenly bodies by pointing out some of the interesting and marvelous phenomena of the universe that are visible with little or no assistance from optical instruments, and indicating means of becoming acquainted with the constellations and the planets. Knowing that an opera-glass is capable of revealing some of the most beautiful sights in the starry dome, and believing that many persons would be glad to learn the fact, he set to work with such an instrument and surveyed all the constellations visible in the latitude of New York, carefully noting everything that it seemed might interest amateur star-gazers. All the objects thus observed have not been included in this book, lest the multiplicity of details should deter or discourage the very readers for whom it was specially written. On the other hand, there is nothing described as visible with an opera-glass or a field-glass which the author has not seen with an instrument of that description, and which any person possessing eyesight of average quality and a competent glass should not be able to discern.

But, in order to lend due interest to the subject, and place it before the reader in a proper light and true perspective, many facts have been stated concerning the objects described, the ascertainment of which has required the aid of powerful telescopes, and to observers with such instruments is reserved the noble pleasure of confirming with their own eyes those wonderful discoveries which the looker with an opera-glass can not hope to behold unless, happily, he should be spurred on to the possession of a telescope. Yet even to glimpse dimly these distant wonders, knowing what a closer view would reveal, is a source of no mean satisfaction, while the celestial phenomena that lie easily within reach of an opera-glass are sufficient to furnish delight and instruction for many an evening.

It should be said that the division of the stars used in this book into the " Stars of Spring," "Stars of Summer," " Stars of Autumn," and "Stars of Winter," is purely arbitrary, and intended only to indicate the seasons when certain constellations are best situated for observation or most conspicuous.

The greater part of the matter composing this volume appeared originally in a series of articles contributed by the author to "The Popular Science Monthly" in 1887-'88. The reception that those articles met with encouraged him to revise and enlarge them for publication in the more permanent form of a book.

G. P. S.
Brooklyn, N. Y., September, 1888.

Table of Contents

Introduction to Stargazing

Stargazing was never more popular than it is now. In every country many excellent telescopes are owned and used, often to very good purpose, by persons who are not practical astronomers, but who wish to see for themselves the marvels of the sky, and who occasionally stumble upon something that is new even to professional star-gazers. Still, it is probably safe to say that hardly one person in a hundred knows the chief stars by name, or can even recognize the principal constellations, much less distinguish the planets from the fixed stars. And of course they know nothing of the intellectual pleasure that accompanies a knowledge of the stars. Modern astronomy is so rapidly and wonderfully linking the earth and the sun together, with the rest of the universe, in the bonds of close physical relationship, that a person of education and general intelligence can offer no good excuse for not knowing where to look for Sirius or Rigel, or the Orion nebula, or the planet Jupiter.

Just as our own world was explored and mapped, so the stars and planets around us are, in a certain sense, falling under the dominion of the restless and resistless mind of man. We have come to possess vested intellectual interests in Mars and Saturn, and in the sun and all his multitude of fellow stars, which nobody can afford to ignore.

A singular proof of popular ignorance of the starry heavens, as well as of popular curiosity concerning any uncommon celestial phenomenon, is furnished by the curious notions prevailing about the planets Mars and Venus.

In August 2003, the planet Mars made its closet approach to Earth in 60,000 years and was a brilliant sight in the night sky. Telescopic observers were treated with the sight of the red deserts and white polar caps of this familiar but still mysterious planet. Yet the erroneous rumor floated the information byways of our own world that Mars would loom as large as the full Moon to our own unaided eye. This rumor was untrue, of course. Only with a telescope with magnification of 100-150x would the planet even approach the apparent size of our full moon, as even the most basic consideration of the facts would show.

And when Venus attracts general attention in the western sky in the early evenings, or the eastern sky in the early mornings, speculation quickly becomes rife about it. As the planet hangs dazzlingly bright, some people appear to think it is a man-made satellite or some less Earthly apparition. This ridiculous notion has been entertained by more than one person of intelligence. And as Venus glowed with increasing splendor in the serene mornings or evenings, she continued to be mistaken for some petty artificial or alien light instead of the magnificent world that she was, sparkling out there in the sunshine like a globe of burnished silver. Yet Venus as an evening star is not so rare a phenomenon that people of intelligence should be surprised at it. Once in every 584 days she reappears at the same place in the sunset sky

"Gem of the crimson-colored even,
Companion of retiring day."

No eye can fail to note her, and as the nearest and most beautiful of the planets it would seem that everybody should be as familiar with her appearance as with the face of a friend. But the popular ignorance of Venus, and Mars, and the other members of the planetary family to which the Earth belongs, is only an index of the denser ignorance concerning the stars. I believe this ignorance is largely due to mere indifference, which, in its turn, arises from a false and pedantic method of presenting astronomy as a creature of mathematical formulae, and a humble handmaiden of the art of navigation. I do not, of course, mean to cast doubt upon the scientific value of technical work in astronomy. The science could not exist without it. Those who have revealed the composition of the sun and stars, and who photograph the heavens as they are, and even reveal phenomena which lie beyond the range of human vision, are the men and women who have set astronomy on its feet as a serious science. But when one sees the depressing and repellent effect that has evidently been produced upon the popular mind by the ordinary methods of presenting astronomy, one can not resist the temptation to utter a vigorous protest, and to declare that this glorious science is not the grinning mathematical skeleton that it has been represented to be.

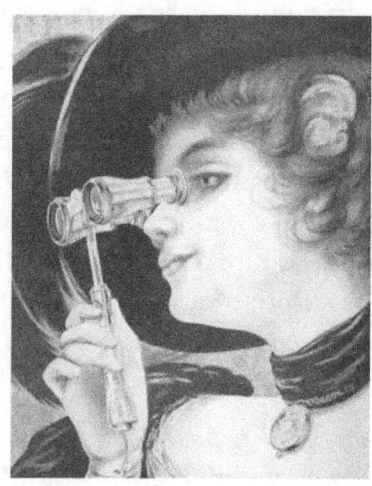

Opera glasses in their natural environment

Perhaps one reason why the average educated man or woman knows so little of the starry heavens is because it is popularly supposed that only the most powerful telescopes and costly instruments of the observatory are capable of dealing with them. No greater mistake could be made. It does not require an optical instrument of any kind, nor much labor, as compared with that expended in the acquirement of some polished accomplishments regarded as indispensable, to give one an acquaintance with the stars and planets which will be not only pleasurable but useful. And with the aid of a simple pair of binoculars, most interesting, gratifying, and, in some instances, scientifically valuable observations may be made in the heavens. I have more than once heard persons who knew nothing about the stars, and probably cared less, utter exclamations of surprise and delight when persuaded to look at certain parts of the sky with binoculars, and thereafter manifest an interest in astronomy of which they would formerly have believed themselves incapable.

Being convinced that whoever will survey the heavens with good binoculars will feel repaid many fold for his time and labor, I have undertaken to point out some of the objects most worthy of attention, and some of the means of making acquaintance with the stars.

Choosing an Instrument

Why Binoculars ?

This book was first written for a reader equipped with an "opera glass", a pair of small side-by-side telescopes used to get close-up views of indoor events. These were simple instruments by today's standards. The objective lenses of an opera glass had diameter of 30-40 mm. And their magnification was perhaps 3-4x: quite reasonable for watching La Boheme from the balcony, but minimally acceptable even for casual astronomy. Today, such an instrument would be considered a toy, not a serious piece of optics.

This book also recommended to its first readers a pair of "field glasses", what we now call binoculars. Few could afford such an instrument a century ago. But today, a good pair of new binoculars can be purchased for less than $150-200, and a quality second-hand pair for considerably less. And the materials and manufacturing techniques used today far exceed the capabilities of the late 19th century. The stars remain the same, but optical technology moves ahead.

Modern hand-held binoculars have objective lenses of at least 35-70 mm aperture, and use internal prisms to allow wide separation of the objectives while keeping the eyepieces close enough for comfortable viewing. The generous aperture of most binoculars helps bring out fainter objects than opera glasses. And magnification of 7-15x makes binoculars more effective in resolving fine detail and rendering a darker background in light polluted skies.

But if you're learning the night sky, why use binoculars at all? Why not a telescope?

Binoculars are less expensive than a telescope, of course. And they are easier and more intuitive to use: you just grab a pair and head outside and start looking.

But for beginning stargazers, the biggest advantage of binoculars is their large field of view. A typical pair lets you see 5 to 8 degrees of sky, about the width of four fingers held at arms length. A telescope lets you see a field of view of less than one degree, which is like looking at the sky through a drinking straw. It's confusing and frustrating for beginners because you can't see very much at once.

Despite their small size, as you'll learn in this book, a modest pair of binoculars lets you see as many as 100,000 stars, hundreds of star clusters and nebula, supernovae remnants, and galaxies. Yes, a telescope is essential for the serious amateur astronomer. But binoculars are the place to start.

Binocular Basics

All binoculars are marked with two key numbers: magnification and aperture. A pair marked "7×50", for example, magnifies 7 times (or 7x) and has objective lenses 50 mm in diameter. The bigger the lenses, the fainter the objects you can see. A pair of 50 mm lenses will collect 50-60 times as much light as your dark-adpated eye.

For astronomy, more aperture is better. A 10×80 pair lets you see fainter objects than a 10X50 pair. The trade-off? Bigger lenses means more weight, and anything larger than 50 mm makes them hard to hold for any length of time.

Higher power means you'll see more detail and a darker background sky. But you'll see a narrower field of view, and it's harder to keep a high-power pair of binoculars steady enough to see fine detail, since the shaking of your arms is also magnified. For hand-held use, magnification of 7-8x is optimum.

A standard pair of Porro-prism binoculars

Mounting binoculars on a camera tripod helps a lot, and lets you use higher power and larger lenses… you get the best of both worlds. Some binoculars come with a pre-drilled hole for mounting. But most smaller binoculars do not, and you have to buy an adapter for your binoculars with the correct ¼-20 mounting hole.

Some binoculars are marked with the size of the field of view, either in degrees or "feet at 1000 yards". This tells you how wide a scene you'll see. For a fixed lens size, higher power means the field of view is lower. So you'll see less sky with 10x50 binoculars, for example, than you will with 7x50 binoculars

Another key measure of binoculars is the "exit pupil", the size of the bright disks of light you see in the eyepieces when you hold the binoculars at arms length. For astronomy, you want the size of these disks to be no larger your eye's pupils when dark-adapted. Otherwise the light collected by the lenses doesn't enter your eye. It's not a problem if this happens: you're just wasting light.

The exit pupil is simply the ratio of aperture to magnification. So a 7×50 pair has an exit pupil of 50/7 = 7 mm (roughly), and a 7×35 pair has a 35/7=5 mm exit pupil.

Under age 30, most people have a dark-adapted exit pupil of 7 mm. But we lose about 1 mm every 10-15 years. At age 50, for example, it may not make sense to use binoculars with an exit pupil larger than 5-6 mm. So if you're older, a pair of 7×35's might be a better choice than a pair of 7×50's. The extra light from the 7x50's won't reach your eye, and they're more expensive.

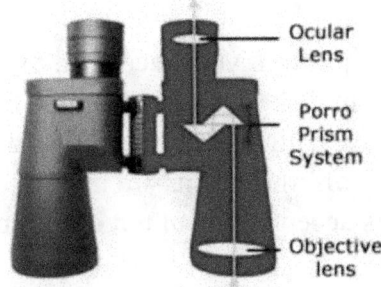

The light path of porro-prism binoculars

How To Choose Binoculars

How do you select a pair of binoculars? Some may disagree, but if you pay less than $75 for a new pair of binoculars, you'll be disappointed with the quality of what you get. On the other hand, almost no one needs to pay more than $300-$400 for an excellent pair. Between $100-300, you'll be spoiled for choice.

Stick with binoculars that use porro-prisms, the classic type of binocular where the objective lens and eyepiece are offset. Binoculars that have a "straight-through" view use roof prisms, and a good pair is expensive. You don't need to pay the premium.

Avoid binoculars with a zoom feature or a built-in camera. They don't make the grade for astronomical use.

When selecting a pair, pick up the binoculars and look at light reflected in the objective lenses. If the lenses have a good anti-reflection coating, they'll appear mostly dark, with a bit of reflected color. If the lenses appear white, or ruby red, don't buy them.

Next, look through the lens at the prisms inside. A good anti-reflection coating shows a colored prism surface. A white surface means no AR coating, which is not recommended.

Now hold the pair away from your face with the eyepieces toward you. Look at the bright disk of the exit pupil. The disk appears round if the prisms use a high-grade glass called BAK-4. If the disk appears squared-off, the prisms are made from lower-grade BK-7 glass, which is acceptable, but not optimum.

If you're near or far sighted, you don't need to wear your glasses when looking through binoculars. You can simply adjust the focus of the binoculars to compensate. But if you have astigmatism, you will need your glasses. Make sure you can see right to the edge of the field of view of the binoculars while wearing your glasses.

Now look through the binoculars, and bring an object into focus at the centre of the field of view. A decent set of optics will hold focus out to the edge of the field. It may not be perfectly focused at the edge, and that's alright. But if the edge of the field is way out of focus or highly distorted, move on to another pair.

What separates a $200 pair of binoculars from a $2000 pair with the same magnification and aperture? The complexity of the AR coatings, the quality of lenses and prisms, and the precision of the lens shape. An expensive pair gives crisp, high-contrast views without distortion right out to the edge of the field. Nice to have, especially for daylight use, but not critical for casual astronomical use.

Image-Stabilized Binoculars

Although they are expensive, image-stabilized (IS) binoculars give stunning low power views of the sky without the dreaded image shake of standard binoculars. And no tripod is required.

Many rave about these technical wonders for astronomy or terrestrial use. In IS binoculars, piezoelectric motion sensors detect pitch and yaw movements. The motion signal feeds into a microprocessor, which initiates image stabilization by controlling a vari-angle prism-- a pair of glass plates joined by flexible bellows. The space between the plates is filled with a silicon-based oil to maximize image deflection.

The motion sensors work in light or total darkness and operate at any orientation, so there are no restrictions on where the binoculars can be pointed... up, down, sideways, anywhere.

When you switch on the IS feature, the image does not "freeze", but rather wanders slowly enough for your eye to follow. And the IS works when you sweep across a field of view, although there is a slight hesitation.

IS binoculars are battery hogs. You can burn through a pair of alkaline batteries in 5 minutes on a cold night. With rechargeables, you might get 2 hours. Of course, you can turn off the IS feature when you're not using it.

Nikon, Canon, and Fujinon, among others, offer some type of image stabilization. Canon models seem to have the widest following among amateur astronomers. A reviewer said of Canon's 10×42 IS binoculars, "These are simply the finest binoculars I have ever used for astronomy". At the time of publishing, these binoculars cost about US$1,200.

If you have dark sky and if you can afford a pair of IS binoculars, they are highly recommended. Mr. Serviss, the author of the first edition of this book in 1888, would have marveled at the technology that makes these binoculars possible.

Canon 12x36 image-stabilized binoculars

The Stars of Spring

Finding Your Way Around The Sky

Having selected your binoculars, the next thing is to find the stars. Of course, you could sweep over the heavens at random on a starry night and see many interesting things, but you would soon tire of such aimless occupation. You must know what you are looking at in order to derive any real pleasure or satisfaction from the sight. It really makes no difference at what time of the year such observations are begun, but for convenience I will suppose that they are begun in the spring. We can then follow the revolution of the heavens through a year, at the end of which, if you are diligent, you will have acquired a competent knowledge of the constellations.

The whole-sky Map 1 represents the appearance of the heavens at midnight on the 1st of March, at eleven o'clock on the 15th of March, at ten o'clock on the 1st of April, at nine o'clock on the 15th of April, and at eight o'clock on the 1st of May.

The reason why a single map can thus be made to show the places of the stars at different hours in different months will be plain upon a little reflection. In consequence of the earth's annual journey around the sun, the whole heavens make one apparent revolution in a year. This revolution, it is clear, must be at the rate of 30 degrees in a month, since the complete circuit comprises 360 degrees.

But, in addition to the annual revolution, there is a daily revolution of the heavens which is caused by the earth's daily rotation upon its axis, and this revolution must, for a similar reason, be performed at the rate of 15 degrees for each of the twenty-four hours. It follows that in two hours of the daily revolution the stars will change their places to the same extent as in one month of the annual revolution. It follows also that, if one could watch the heavens throughout the

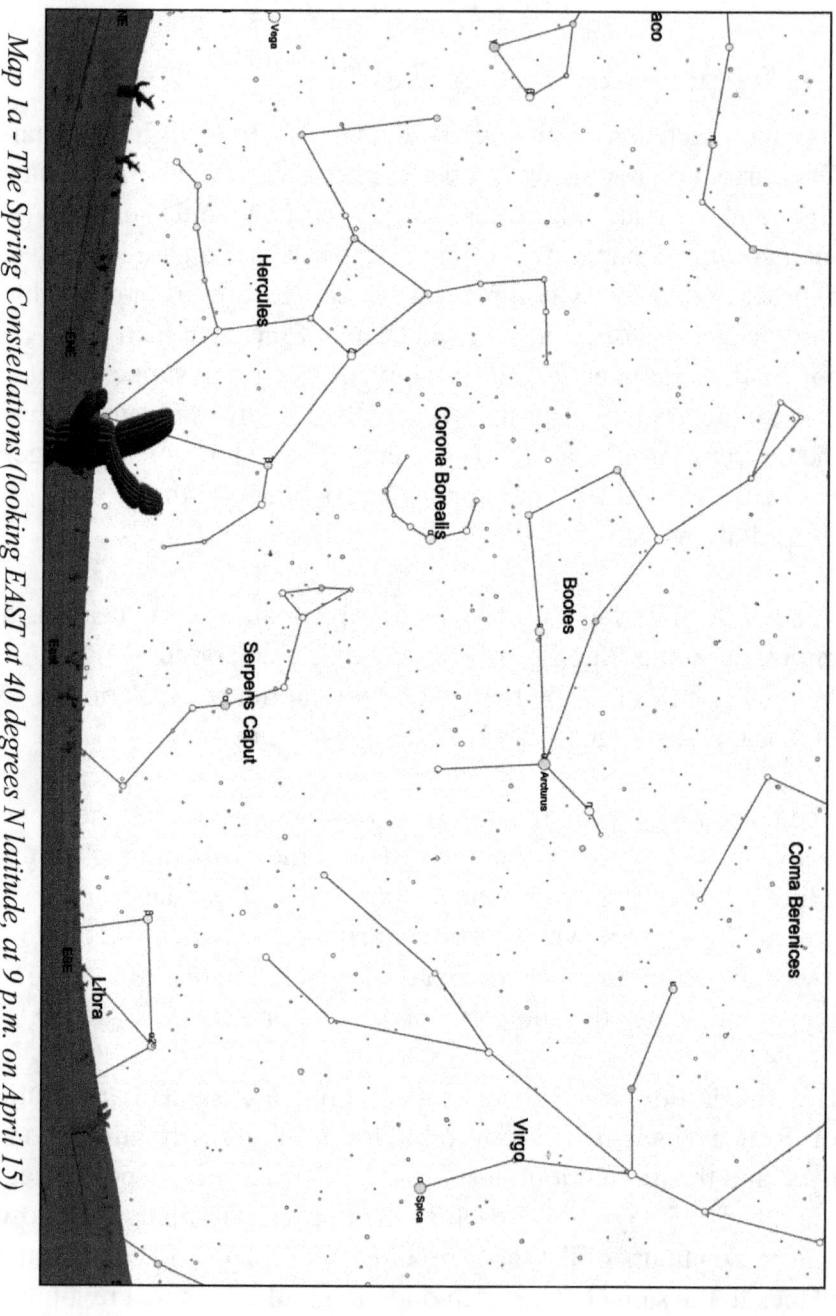

Map 1a - The Spring Constellations (looking EAST at 40 degrees N latitude, at 9 p.m. on April 15)

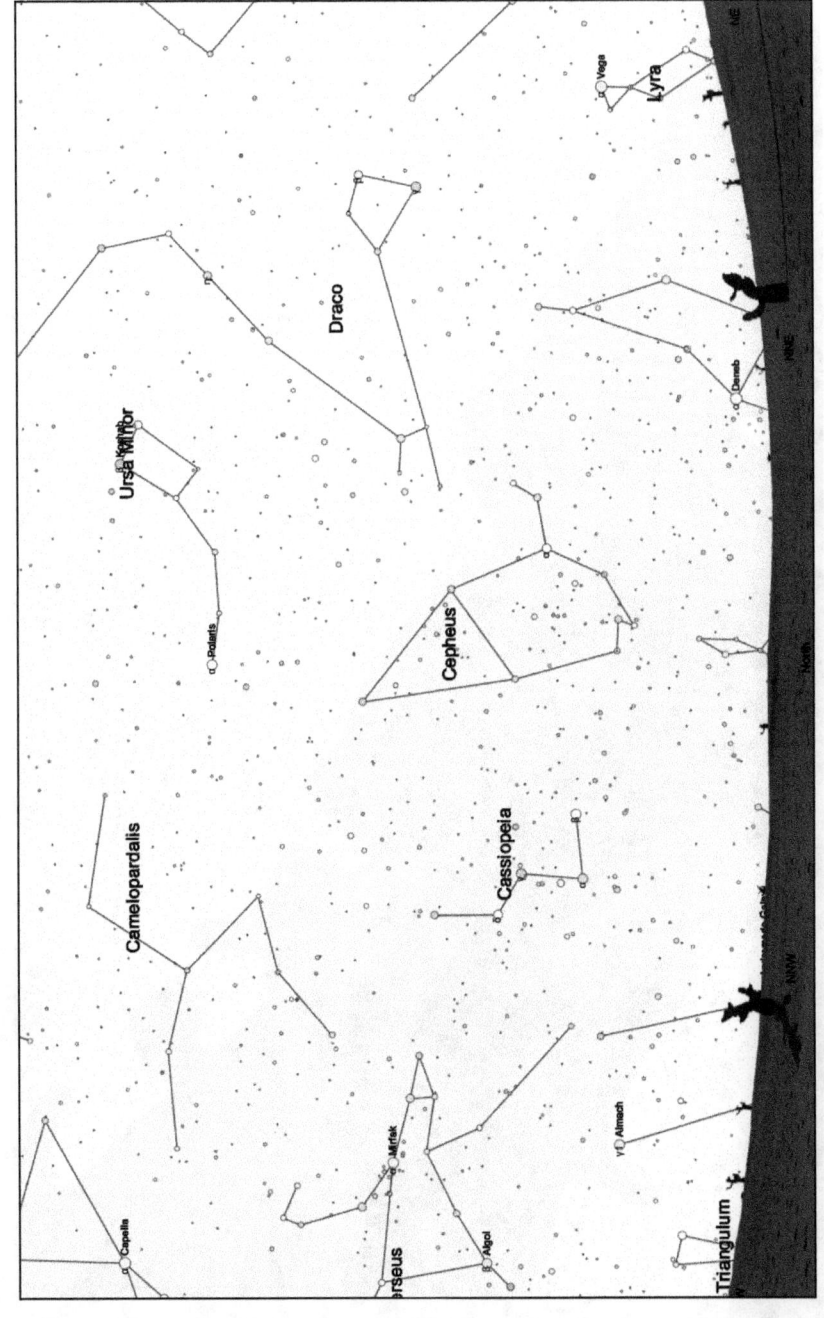

Map 1b - The Spring Constellations (looking NORTH at 40 degrees N latitude, at 9 p.m. on April 15)

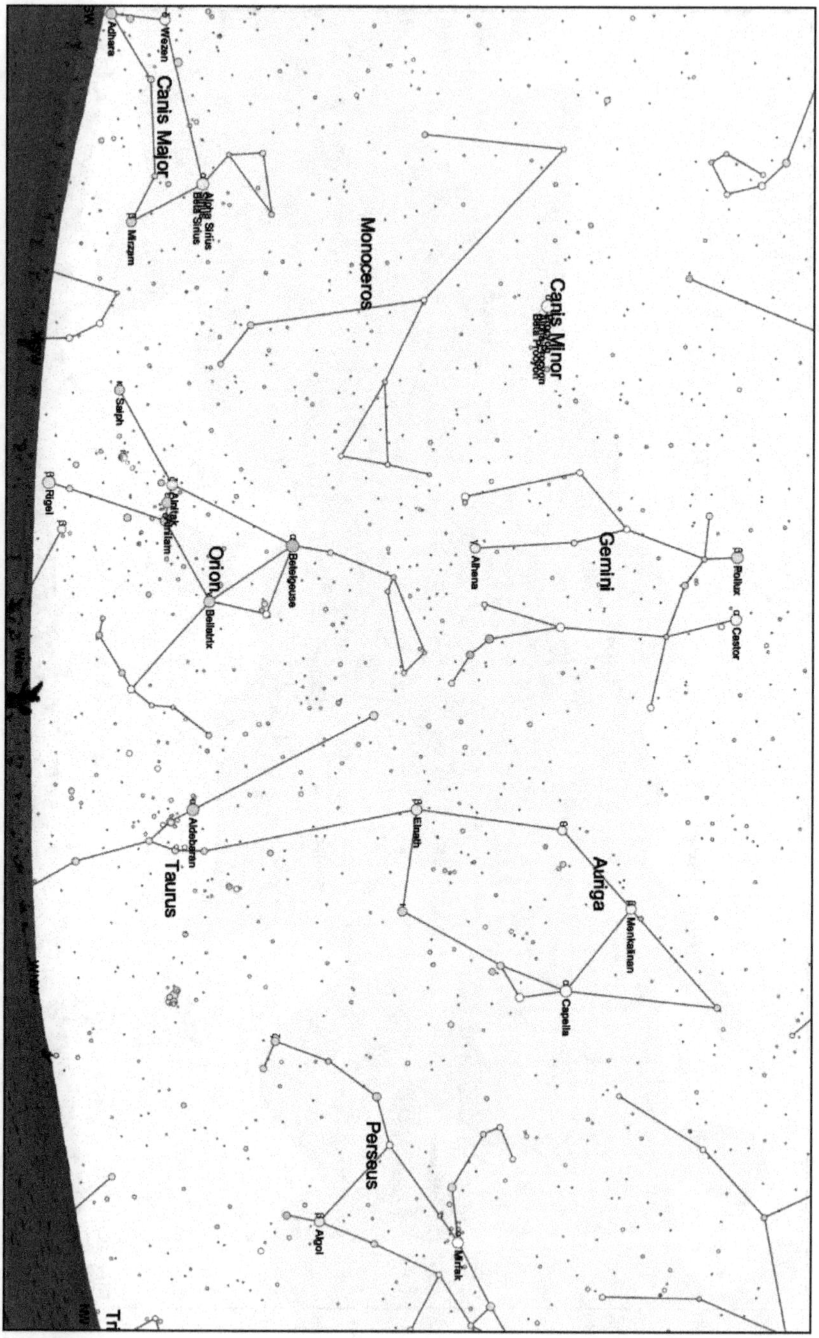

Map 1c - The Spring Constellations (looking WEST at 40 degrees N latitude, at 9 p.m. on April 15)

Canis Major

Monoceros

Canis Minor

Gemini

Orion

Taurus

Auriga

Perseus

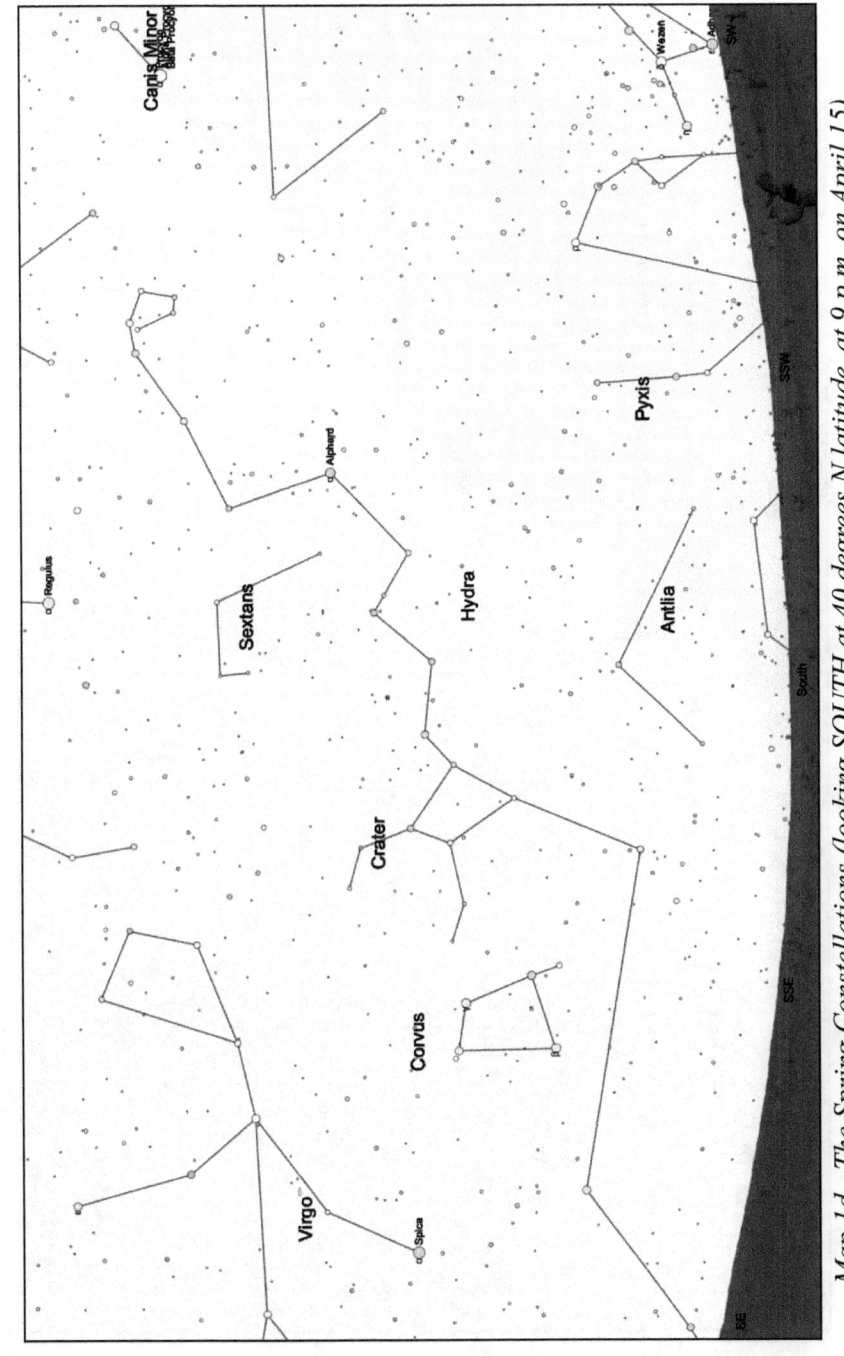

Map 1d - The Spring Constellations (looking SOUTH at 40 degrees N latitude, at 9 p.m. on April 15)

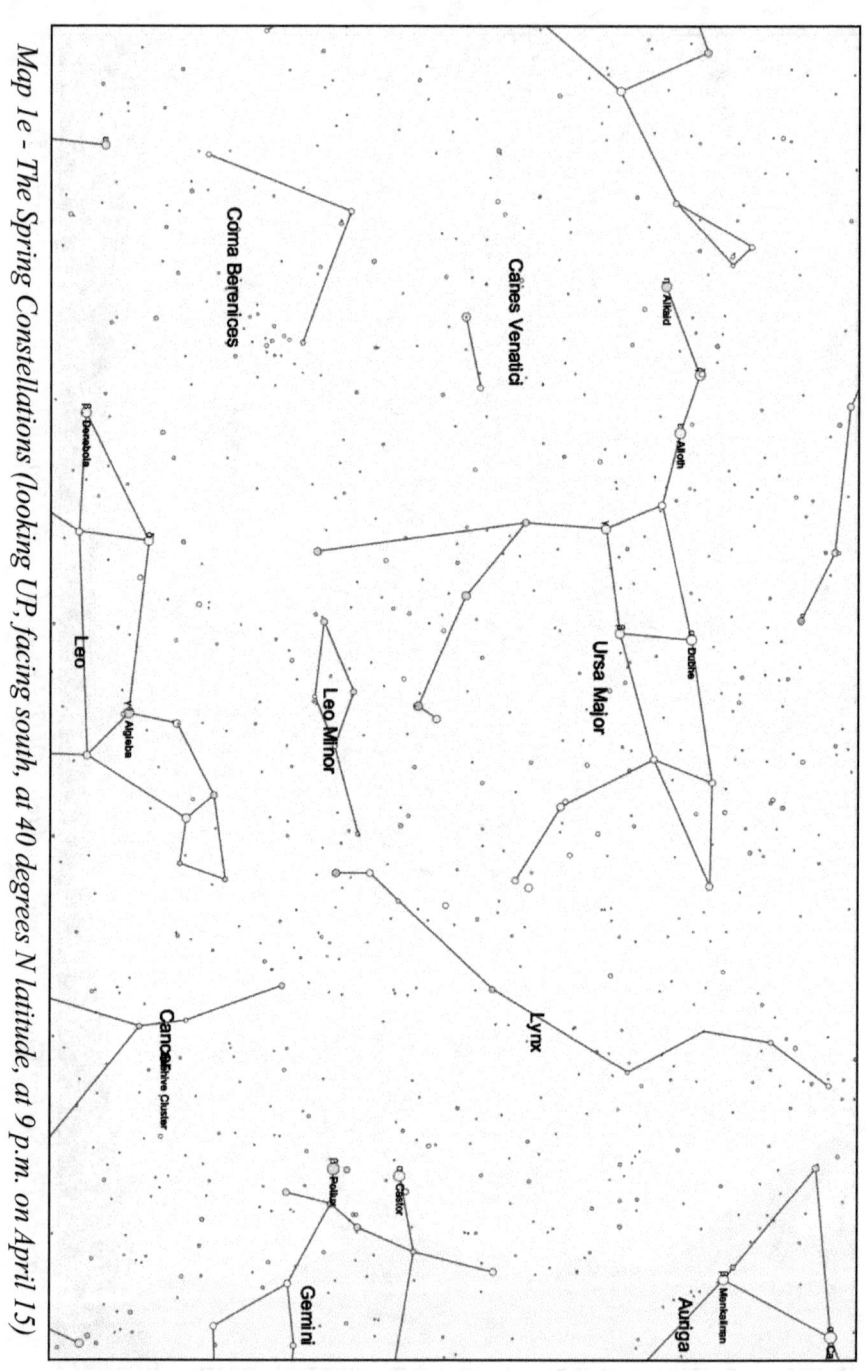

Map 1e - The Spring Constellations (looking UP, facing south, at 40 degrees N latitude, at 9 p.m. on April 15)

whole twenty-four hours, and not be interrupted by daylight, he would see the complete circuit of the stars just as he would do if, for a year, he should look at the heavens at a particular hour every night.

Suppose that at nine o'clock on the 1st of June we see the star Spica on the meridian (when it's at its highest point in the sky); in consequence of the rotation of the earth, two hours later, or at eleven o'clock, Spica will have moved 30 degrees west of the meridian. But that is just the position which Spica would occupy at nine o'clock on the 1st of July, for in one month (supposing a month to be accurately the twelfth part of a year) the stars shift their places 30 degrees toward the west. If, then, we should make a map of the stars for nine o'clock on the 1st of July, it would answer just as well for eleven o'clock on the 1st of June, or for seven o'clock on the 1st of August.

Leo, The Lion

The center of the map is the zenith, the point directly overhead. The reader must now exercise his imagination a little, for it is impossible to represent the true appearance of the concave of the heavens on flat paper. Holding the map over your head, with the points marked East, West, North, and South in their proper places, conceive of it as shaped like the inside of an open umbrella, the edge all around extending clear down to the horizon. Suppose you are facing the south, then you will see, up near the zenith, the constellation of Leo, which can be readily recognized on the map by six stars that mark out the figure of a sickle standing upright on its handle. The large star in the bottom of the handle is Regulus. Having fixed the appearance and situation of this constellation in your mind, go out-of-doors, face the south, and try to find the constellation in the sky. With a little application you will be sure to succeed.

Using Leo as a basis of operations, your conquest of the sky will now proceed more rapidly. By reference to the map you will be able to recognize the twin stars of Gemini, southwest of the zenith and high up; the brilliant lone star, Procyon, south of Gemini; the dazzling Sirius, flashing low down in the southwest; Orion, with all his brilliant stars, blazing in the west; red Aldebaran and the Pleiades off to his right; and Capella, bright as a diamond, high up above Orion, toward the north. In the southeast you will recognize the quadrilateral of Corvus, with the remarkably white star Spica glittering east of it.

Next face the north. If you are not just sure where north is, try a pocket-compass. This advice is by no means unnecessary, for there are many intelligent persons who are unable to find true north within many degrees, though standing on their own doorstep. Having found the north point as near as you can, look upward about forty degrees from the horizon (or the number of degrees equivalent to your latitude), and you will see the lone twinkler called the north or pole star. If you live at forty degrees north latitude, the pole star will lie forty degrees above the horizon, a little less than half-way from the horizon to the zenith.

By the aid of the map, again, you will be able to find, high up in the northeast, near the zenith, the large dipper-shaped figure in Ursa Major, and, when you have once noticed that the two stars in the outer edge of the bowl of the Dipper point almost directly to the pole-star, you will have an unfailing means of picking out the latter star hereafter.* Continuing the curve of the Dipper-handle, in the northeast, your eye will be led to a bright reddish star, which is

_____

* Let the reader remember that the distance between the two stars in the brim of the bowl of the Big Dipper is about 10 degrees, and he will have a measuring stick that he can apply in estimating other distances in the heavens.

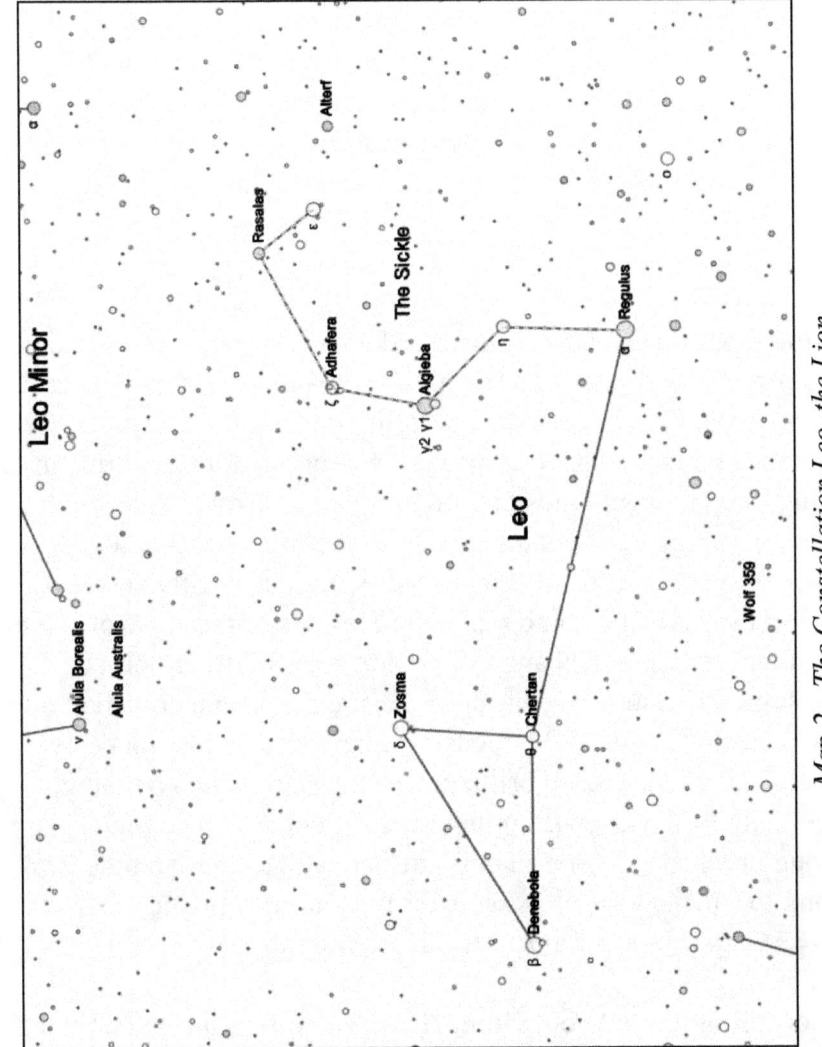

Map 2 - The Constellation Leo, the Lion

Arcturus, in the constellation Bootes. The arc extends further to the remarkable white star Spica.

In the same way you will be able to find the constellations Cassiopeia, Cepheus, Draco, and Perseus. Don't expect to accomplish it all in an hour. You may have to devote two or three evenings to such observation, and make many trips indoors to consult the map, before you have mastered the subject; but when you have done it you will feel amply repaid for your exertions, and you will have made for yourself silent friends that will beam kindly upon you, like old neighbors, on whatever side of the world you may wander.

Having fixed the general outlines and location of the constellations in your mind, and learned to recognize the chief stars, take your binoculars and begin with the constellation Leo and the star Regulus. Contrive to have some convenient rest for your arms in holding the glass, and thus obtain not only comfort but steadiness of vision. A lazy-back chair makes a capital observing seat. Be very particular, too, to get a sharp focus. Remember that no two persons' eyes are alike, and that even the eyes of the same observer occasionally require a change. In looking for a difficult object, I have sometimes suddenly brought the sought-for phenomenon into view by a slight turn of the focusing screw. You will at once be gratified by the increased brilliancy of the star as seen by the glass. If the night is clear, it will glow like a diamond. Yet Regulus, although ranked as a first-magnitude star, and of great repute among the ancient astrologers, is far inferior in brilliancy to such stars as Capella and Arcturus, to say nothing of Sirius.

By consulting Map 2 you will next be able to find the celebrated star bearing the name of Algieba, also called Gamma (Υ) Leonis. If you had a telescope, you would see this star as a close and beautiful double, of contrasted colors. But in binoculars, it appears as an optical double. You cannot fail to see a small star near it. You will be struck by the surprising change of color in turning from Regulus to Gamma, as the former is white and the latter deep yellow. It will be well to look first at one and then at the other, several times, for this is a good instance of what you will meet with many times in your future surveys of the heavens: a striking contrast of color in neighboring stars.

One can thus comprehend that there is more than one sense in which to understand the Scriptural declaration that "one star differeth from another in glory." The radiant point of the famous November meteors, called the Leonids, is near Gamma.

Turn next to the star in Leo called Adhafera. If your glass is 40 mm or more, and your eye keen, you will easily see three tiny companion stars keeping company with Adhafera, two on the southeast, and one, much closer, toward the north. The nearest of the two on the south is faint, being only between the eighth and ninth magnitude, and will probably severely test your powers of vision. Next look at the star ε (Epsilon) Leonis, and you will find near it two seventh-magnitude companions, making a beautiful little triangle.

Away at the eastern end of the constellation, in the tail of the imaginary Lion, upon whose breast shines Regulus, is the star β (Beta) Leonis, also called Denebola. It is almost as bright as its leader, Regulus, and you will probably be able to catch a tinge of blue in its rays. South of Denebola, at a distance of nineteen minutes of arc, or somewhat more than half the apparent diameter of the moon, you will see a little star of the sixth magnitude, which is one of the several "companions" for which Denebola is celebrated. There is another star of the eighth magnitude in the same direction from Denebola, but at a distance of less than five minutes, and this you may be able to glimpse with binoculars under favorable conditions. I have seen it well with binoculars of 40 mm aperture, and a magnifying power of seven times. But it

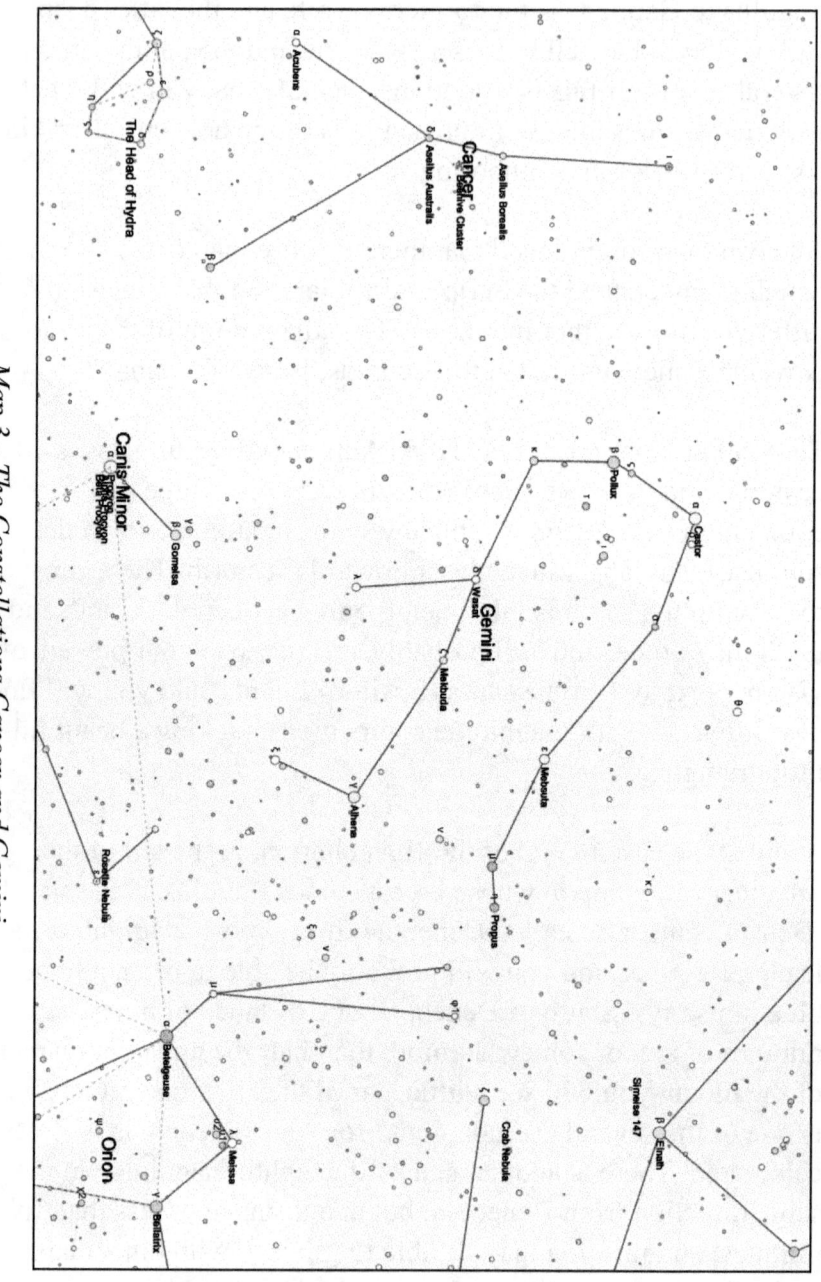

Map 3 - The Constellations Cancer and Gemini

requires an experienced eye and steady vision to catch this shy twinkler.

When looking for a faint and difficult object, the plan pursued by telescopists is to avert the eye from the precise point upon which the attention is fixed, in order to bring a more sensitive part of the retina into play. Look toward the edge of the field of view, while the object you are seeking is in the center, and then, if it can be seen at all with your glass, you will catch sight of it, as it were, out of the corner of your eye. The effect of seeing a faint star in this way, in the neighborhood of a large one, whose rays hide it from direct vision, is sometimes very amusing. The little star seems to dart out into view as through a curtain, perfectly distinct, though as immeasurably minute as the point of a needle. But the instant you direct your eyes straight at it, *presto!* it is gone. And so it will dodge in and out of sight as often as you turn your eyes.

If you will sweep carefully over the whole extent of Leo, you will be impressed with the power of your optics to bring into sight many faint stars in regions that seem barren to the naked eye. Binoculars of just 30-35 mm aperture will show twenty times as many stars as the naked eye can see.

A word about the "Lion" which this constellation is supposed to represent. It requires a vivid imagination to perceive the outlines of the celestial king of beasts among the stars, and yet somebody taught the people of ancient India and the old Egyptians to see him there, and there he has remained since the dawn of history. Modern astronomers strike the historical image out of their charts, together with all the picturesque multitude of beasts and birds and men and women that bear him company, but they cannot altogether banish the old names, and, practically, the old outlines of the constellations are retained, and always will be retained. The Lion is the most conspicuous figure in the celebrated zodiac; and, indeed, there is evidence that before the story of Hercules and his labors was told this lion was already imagined shining among the stars. It was characteristic of the Greeks that they seized him for their own, and tried to rob him of his real antiquity by pretending that Zeus had placed him among the stars in commemoration of Hercules' victory over the Nemsean lion. In the Hebrew zodiac Leo represented the Lion of Judah. It was thus always a lion that the ancients thought they saw in this constellation.

In the old star maps the Lion is represented as in the act of springing upon his prey. His face is to the west, and the star Regulus is in his heart. The sickle-shaped figure covers his breast and head, Algieba being in the shoulder, Adhafera in the mane of the neck, and Rasalas and Epsilon Leonis in the cheek. The fore-paws are drawn up to the breast and represented by the stars around Subra. Denebola is in the tuft of the tail. The hind-legs are extended downward at full length, in the act of springing. Starting from the star Zosma in the hip, the row consisting of Chertan, Iota (ι), Tau (τ), and Upsilon (υ) Leonis, shows the line of the hind-legs.

Leo had an unsavory reputation among the ancients because of his supposed influence upon the weather. The greatest heat of summer was felt when the sun was in this constellation:

> *"Most scorching is the chariot of the Sun,*
> *And waving spikes no longer hide the furrows*
> *When he begins to travel with the Lion."*

Cancer, The Crab; The Zodiac

Looking now westward from the Sickle of Leo, at a distance about equal to twice the length of the Sickle, your eye will be caught by a small silvery spot in the sky lying nearly between two rather faint stars. This is the famous Praesepe, or "Manger", in the center of the constellation Cancer. The two stars on either side of it are called the Aselli, or the Ass's Colts, and the imagination of the ancients pictured them feeding from their silver manger. Turn your glass upon the Manger and you will see that it consists of a crowd of little stars, so small and numerous that you will probably not undertake to count them. Galileo has left a delightful description of his surprise and gratification when he aimed his telescope at this curious cluster and other similar aggregations of stars and discovered what they really were. Using his best instrument, he was able to count thirty-six stars in the Manger. The Manger was a famous weather-sign in olden times, and Aratus, in his "Diosemia," advises his readers to:

> " . . . watch the Manger : like a little mist
> Far north in Cancer's territory it floats.
> Its confines are two faintly glimmering stars ;
> These are two asses that a manger parts,
> Which suddenly, when all the sky is clear,
> Sometimes quite vanishes, and the two stars
> Seem to have closer moved their sundered orbs.
> No feeble tempest then will soak the lea ;
> A murky manger with both stars
> Shining unaltered is a sign of rain."

Like other old weather-saws, this probably possesses a gleam of sense, for it is only when the atmosphere is perfectly transparent that the Manger can be clearly seen; when the air is thick with mist, the harbinger of coming storm, it fades from sight. Light pollution of the type seen in moderate to large cities renders the Manger difficult or impossible to see. The manger was also cataloged by the comet hunter Charles Messier. He gave it the 44th position in his famous list; for this reason, the manger is also called M44 (or Messier 44). You will meet more Messier objects in the coming pages.

The constellation Cancer, or the Crab, was represented by the Egyptians under the figure of a Scarab Beetle. The observer will probably think that it is as easy to see a beetle as a crab there. Cancer, like Leo, is one of the twelve constellations of the Zodiac, the name applied to the imaginary zone 16 degrees wide and extending completely around the heavens, the center of which is the ecliptic or annual path of the sun. The names of these zodiacal constellations, in their order, beginning at the west and counting round the circle, are: Aries, Taurus, Gemini, Cancer, Leo, Virgo, Libra, Scorpio, Sagittarius, Capricornus, Aquarius, and Pisces. Cancer has given its name to the circle called the Tropic of Cancer, which indicates the greatest northerly declination of the sun in summer, and which he attains on the 21st or 22d of June. But, in consequence of the precession of the equinoxes, all of the zodiacal constellations are continually shifting toward the east, and Cancer has passed away from the place of the summer solstice, which is now to be found in Gemini.

Hydra, The Serpent

Below the Manger, a little way toward the south, your eye will be caught by a group of four or five stars of about the same brightness as the Aselli. This marks the head of Hydra, and your glass will show a striking and beautiful geometrical arrangement of the stars composing it. Hydra is a very long constellation, and trending southward and eastward from the head it passes underneath Leo, and, sweeping pretty close down to the horizon, winds away under Corvus, the tail reaching to the eastern horizon. The length of this sky serpent is about 100 degrees. Its stars are all faint, except Alphard, or the Hydra's Heart, a second-magnitude star, remarkable for its lonely situation, southwest of Regulus. A line from Gamma Leonis through Regulus points it out. It is worth looking at with the glass on account of its rich orange tint.

Hydra is fabled to be the hundred-headed monster that was slain by Hercules. It must be confessed that there is nothing very monstrous about it now except its length. The most timid can look upon it without suspecting its grisly origin.

Gemini, The Twins

Coming back to the Manger as a starting-point, look well up to the north and west, and at a distance somewhat less than that between Regulus and the Manger you will see a pair of first-magnitude stars, which are the celebrated Twins, from which the constellation Gemini takes its name. The star marked α (Alpha) in the map is Castor, and the star marked β (Beta) is Pollux. A classical reader need not be reminded of the romantic origin of these names.

A sharp contrast in the color of Castor and Pollux comes out as soon as your glass is turned upon them. Castor is white, while Pollux is deep yellow. Castor is a celebrated double star, but its components are far too close to be separated with binoculars. You will be at once interested by the singular cortege of small stars by which both Castor and Pollux are surrounded. These little attendant stars, for such they seem, are arrayed in symmetrical groups pairs, triangles, and other figures which, it seems difficult to believe, could be unintentional, although it would be still more difficult to suggest any reason why they should be arranged in that way.

Map 3 will show you the position of the principal stars of the constellation. Castor and Pollux are in the heads of the Twins, while the row of stars shown in the map ξ (Xi), γ (Gamma) or Alhena, μ(Mu), ν (Nu) Gem, and η (Eta) or Propus marks their feet, which are dipped in the edge of the Milky-Way. One can spend a profitable and pleasurable half-hour in exploring the wonders of Gemini. The whole constellation, from head to foot, is gemmed with stars which escape the naked eye, but it sparkles like a bead-spangled garment when viewed with binoculars.

Owing to the presence of the Milky-Way, the spectacle around the feet of the Twins is particularly magnificent. And here the

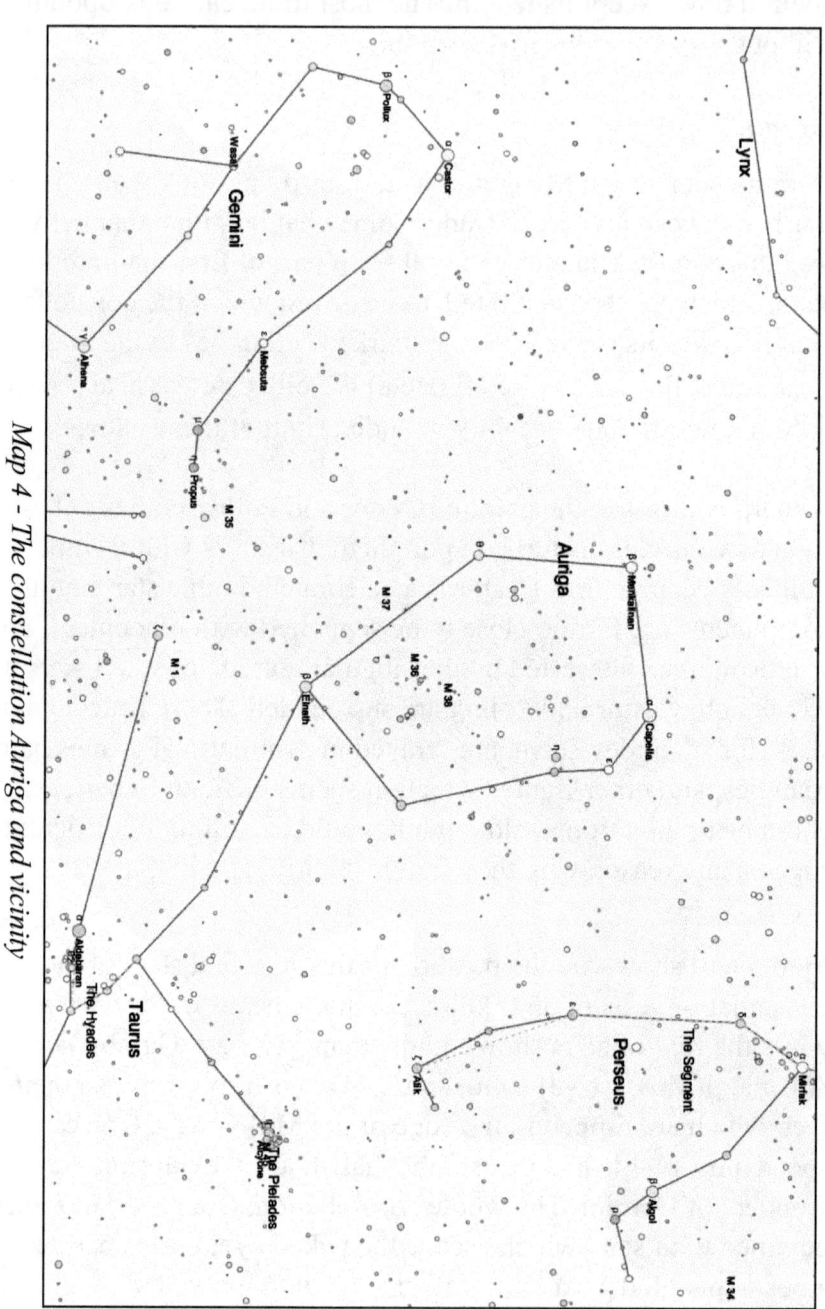

Map 4 - The constellation Auriga and vicinity

possessor of binoculars can get a fine view of a celebrated star-cluster known in the catalogues as M35. It is situated a little distance northwest of the star Propus, and is visible to the naked eye, on a clear, moonless night, as a nebulous speck. With a good glass you will see two wonderful streams of little stars starting, one from Propus and the other from Nu, and running parallel toward the northwest; M35 is situated between these star-streams.

The stars in the cluster are so closely aggregated that you will be able to clearly separate only the outlying ones. The general aspect is like that of a piece of frosted silver over which a twinkling light is playing. The splendor of this starry congregation, viewed with a powerful telescope, may be guessed at from Admiral Smyth's picturesque description: *"It presents a gorgeous field of stars, from the ninth to the sixteenth magnitude, but with the center of the mass less rich than the rest. From the small stars being inclined to form curves of three or four, and often with a large one at the root of the curve, it somewhat reminds one of the bursting of a sky-rocket."* And Webb adds that there is an *"elegant festoon near the center, starting with a reddish star."*

No one can gaze upon this marvelous phenomenon, even with the comparatively low powers of binoculars, and reflect that all these swarming dots of light are really suns, without a stunning sense of the immensity of the material universe.

It is an interesting fact that the summer solstice, or the point which the sun occupies when it attains its greatest northerly declination, on the longest day of the year, is close by this great cluster in Gemini. In the glare of the sunshine those swarming stars are then concealed from our sight, but with the mind's eye we can look past and beyond our sun, across the incomprehensible chasm of space, and behold them still shining, their commingled rays making our own sun seem but a lonely wanderer in the expanse of the universe.

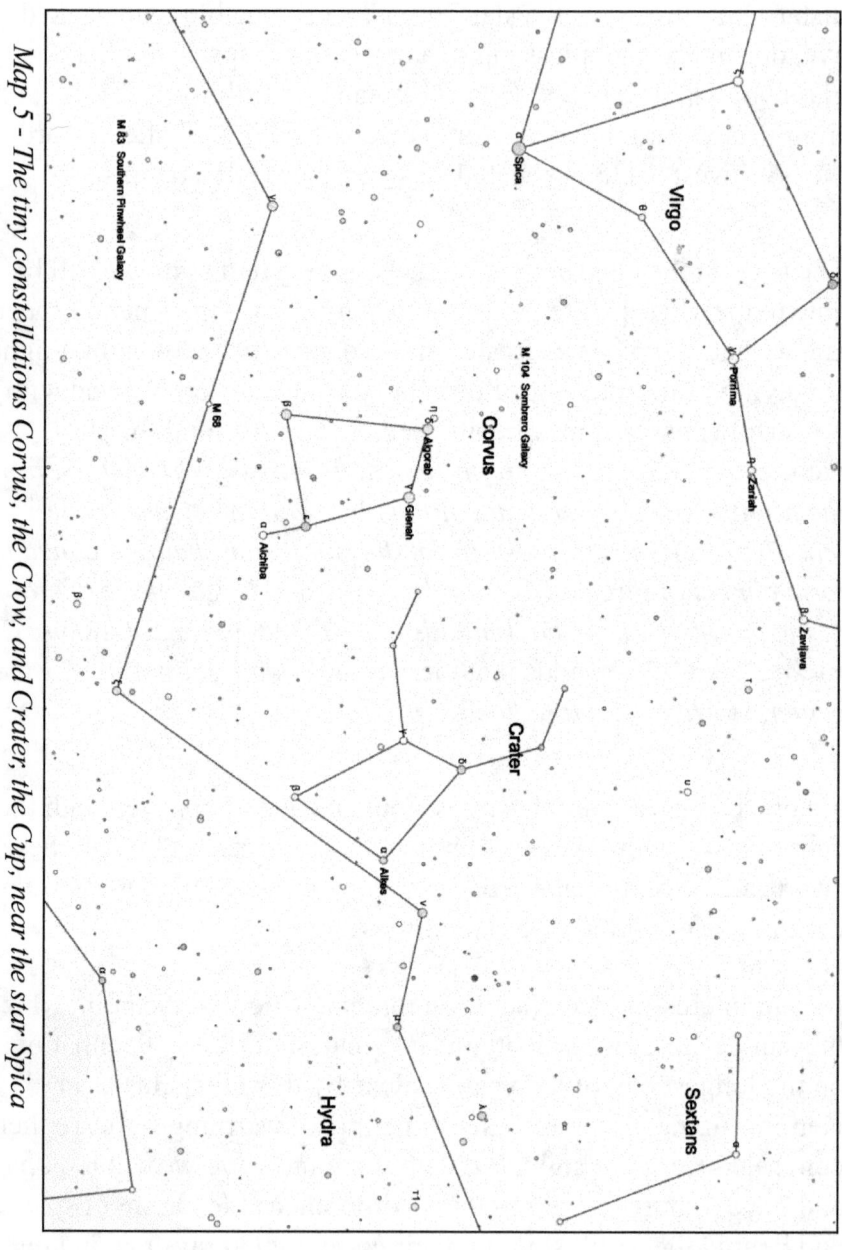

Map 5 - *The tiny constellations Corvus, the Crow, and Crater, the Cup, near the star Spica*

38

It was only a short distance southwest of this cluster that one of the most celebrated discoveries in astronomy was made. There, on the evening of March 13, 1781, William Herschel observed a star whose singular aspect led him to put a higher magnifying power on his telescope. The higher power showed that the object was not a star but a planet, or a comet, as Herschel at first supposed. It was the planet Uranus, whose discovery *"at one stroke doubled the breadth of the sun's dominions."*

The constellation of Gemini, as the names of its two chief stars indicate, had its origin in the classic story of the twin sons of Jupiter (or Zeus) and Leda:

> *"Fair Leda's twins, in time to stars decreed,*
> *One fought on foot, one curbed the fiery steed."*

Castor and Pollux were regarded by both the Greeks and the Romans as the patrons of navigation, and this fact crops out very curiously in the adventures of St. Paul. After his disastrous shipwreck on the island of Melita he embarked again on a more prosperous voyage in a ship bearing the name of these very brothers. "And after three months," writes the celebrated apostle (Acts xxviii, 11) "we departed in a ship of Alexandria, which had wintered in the isle, whose sign was Castor and Pollux." We may be certain that Paul was acquainted with the constellation of Gemini, not only because he was skilled in the learning of his times, but because, in his speech on Mars Hill, he quoted a line from the opening stanzas of the poet Aratus' "Phenomena", a poem in which the constellations are described.

Canis Minor, The "Little Dog"

Map 3 will enable you next to find Procyon, or the Little Dog-Star, more than twenty degrees south of Castor and Pollux, and almost directly below the Manger. This star will interest you by its golden-yellow color and its brightness, although it is far inferior in the latter respect to Sirius, or the Great Dog-Star, which you will see flashing splendidly far down beneath Procyon in the southwest. About four degrees northwest of Procyon is a third-magnitude star, called Gomeiza, and the glass will show you two small stars which make a right-angled triangle with it, the nearer one being remarkable for its ruddy color.

Procyon is especially interesting because it is attended by an invisible star, which was first perceived by its effect of its attraction upon Procyon. It is a curious fact that both of the so-called Dog-Stars are thus attended by obscure or dusky companion-stars, which, notwithstanding their lack of luminosity, are of great mass.

As in the case of Sirius, a large telescope has brought the mysterious attendant of Procyon into view. Almost half a century ago the famous German scientist Bessel announced his conclusion that both Sirius and Procyon were binary systems, consisting each of a visible and an invisible star. He calculated the probable period of revolution, and found it to be, in each case, approximately fifty years. Sixteen years after Bessel's death, one of Alvan Clark's unrivaled telescopes at last revealed the strange companion of Sirius, a huge body, half as massive as the giant Dog-Star itself, but ten thousand times less brilliant, and more recent observations have shown that its period of revolution is within six or seven months of the fifty years assigned by Bessel. The faint but massive companion stars to Sirius and Procyon are now known to be a type of dense stellar cinder, the remains of a mid-sized star that has settled into its final phase of life as a white dwarf.

The mythological history of Canis Minor is somewhat obscure. According to various accounts it represents one of the goddess Diana's hunting-dogs, one of Orion's hounds, the Egyptian dog-headed god Anubis, and one of the dogs that devoured their master Actseon after Diana had turned him into a stag.

The reader will wonder all the more at these legends after he has succeeded in picking out the modest Little Dog in the sky.

Auriga

Sirius, Orion, Aldebaran, and the Pleiades, all of which you will perceive in the west and southwest, are generally too much involved in the mists of the horizon to be seen to the best advantage at this season, although it will pay you to take a look through the glass at Sirius.

But the splendid star Capella, in the constellation Auriga, may claim a moment's attention. You will find it high up in the northwest, half-way between Orion and the pole-star, and to the right of the Twins. It has no rival near, and its creamy-white light makes it one of the most beautiful as well as one of the most brilliant stars in the heavens. Its constitution, as revealed by the spectroscope, resembles that of our sun, but the sun would make but a sorry figure if removed to the distance of this giant star. About seven and a half degrees above Capella, and a little to the left, you will see a second-magnitude star called Menkalinan. Two and a half times as far to the left, or south, in the direction of Orion, is another star of equal brightness to Menkalinan. This is Alnath, and marks the place where the foot of Auriga, or the Charioteer, rests upon the point of the horn of Taurus.

Capella, Menkalinan, and Alnath make a long triangle which covers the central part of Auriga. In clear, dark sky, the naked eye shows two or three misty-looking spots within this triangle, one to the right of Alnath, one in the upper or eastern part of the constellation, near the third-magnitude star Theta (θ), and another on a line drawn from Capella to Alnath, but much nearer to

Capella. Turn your glass upon these spots, and you will be delighted by the beauty of the little stars to whose united rays they are due. These are the rich star clusters marked M38, M37, and M36.

Alnath has around it some very remarkable rows of small stars, and the whole constellation of Auriga, like that of Gemini, glitters with star-dust, for the Milky-Way runs directly through it.

The mythology of Auriga is not clear, but the ancients seem to have been of one mind in regarding the constellation as representing the figure of a man carrying a goat and her two kids in his arms. Auriga was also looked upon as a beneficent constellation, and the goat and kids were believed to be on the watch to rescue shipwrecked sailors. As Capella, which represents the fabled goat, shines nearly overhead in winter, and would ordinarily be the first bright star to beam down through the breaking clouds of a storm at that season, it is not difficult to imagine how it got its reputation as the seaman's friend.

If you wish to to exercise your fancy by trying to trace the outlines of this figure, you will find the head of Auriga marked by the star Delta (δ) and the little group near it. Capella, in the heart of the Goat, is just below his left shoulder, and Menkalinan marks his right shoulder. Alnath is in his right foot, and Iota (ι) in his left foot. The stars Epsilon (ε), Zeta (ζ), Eta (η), and Lambda (λ) shine in the kids which lie in Auriga's lap. The faint stars scattered over the eastern part of the constellation are sometimes represented as forming a whip with many lashes, which the giant flourishes with his right hand.

Let us turn back to Denebola in the Lion's Tail. Now glance from it down into the southeast, and you will see a brilliant star flashing well above the horizon. This is Spica, the brightest star of Virgo, and it is marked on our circular map. Then look into the northwest, and at about the same distance from Denebola, but higher above the horizon than Spica, you will catch the sparkling of a large, reddish star. It is Arcturus in Bootes. The three, Denebola, Spica, and Arcturus, mark the corners of a great equilateral triangle. Nearly on a line between Denebola and Arcturus, and somewhat nearer to the former, you will perceive a curious twinkling, as if gossamers spangled with dew-drops were entangled there. One might think the old woman of the nursery rhyme who went to sweep the cobwebs out of the sky had skipped this corner, or else that its delicate beauty had preserved it even from her best instincts. This is the little constellation called Coma Berenices (Berenice's Hair). Your binoculars will enable you to count twenty or thirty of the largest stars composing this cluster, which are arranged, as so often happens, with a striking appearance of geometrical design.

The constellation has a very romantic history. It is related that the young Queen Berenice, when her husband was called away to the wars, vowed to sacrifice her beautiful tresses to Venus if he returned victorious over his enemies. He did return home in triumph, and Berenice, true to her vow, cut off her hair and bore it to the Temple of Venus. But the same night it disappeared. The king was furious, and the queen wept bitterly over the loss. There is no telling what might have happened to the guardians of the temple, had not a celebrated astronomer named Conon led the young king and queen aside in the evening and showed them the missing locks shining transfigured in the sky. He assured them that Venus had placed Berenice's lustrous ringlets among the stars, and, as they were not skilled in celestial lore, they were quite ready to believe that the silvery swarm they saw near Arcturus had never been there before. And so for centuries the world has recognized the constellation of Berenice's Hair.

Look next at Corvus and Crater, the Crow and the Cup, two little constellations which you will discover on the circular map, and of which we give a separate representation in Map 5. You will find that the stars δ (Delta, or Gienah) and η (Eta), in the upper left-hand corner of the quadrilateral figure of Corvus, make a striking appearance. The little star Zeta (ζ) is a very pretty double for binoculars. There is a very faint pair of stars close below and to the right of Beta (β). This forms a severe test. Only a good set of binoculars will show both, one being considerably fainter than the other. Crater is worth sweeping over for the pretty combinations of stars to be found in it.

You will observe that the interminable Hydra extends his lengthening coils along under both of the constellations. In fact, both the Cup and the Crow are represented as standing upon the huge serpent. The outlines of a cup are tolerably well indicated by the stars included under the name Crater, but the constellation of the Crow might as well have borne any other name so far as any traceable likeness is concerned. One of the legends concerning Corvus says it is the daughter of the King of Phocis, who was transformed into a crow to escape the pursuit of Neptune. She is certainly safe in her present guise.

Arcturus and Spica, and their companions, may be left for observation to a more convenient season, when, having risen higher, they can be studied to better advantage. It will be well, however, to merely glance at them with the glass in order to note the great difference of color Spica being brilliantly white and Arcturus almost red.

We will now turn to the north. You have already been told how to find the pole-star. Look at it with your glass. The pole-star is a famous double, but its minute companion can only be seen with a telescope. As so often happens, however, it has by chance alignment another companion for binoculars, and this latter is sufficiently close and small to make an interesting test for an inexperienced observer armed with a glass of small power. It must be looked for close to the rays of the large star. It is of the seventh magnitude. With larger binoculars, several smaller companions may be seen, and a very excellent glass may show an 8.5-magnitude star almost hidden in the rays of the seventh- magnitude companion.

With the aid of Map 6 find in Ursa Minor, which is the constellation to which the pole-star belongs, the star Beta (β), which is also called Kocab (the star marked Alpha (α) in the map is the pole-star, Polaris). Kocab has a pair of faint stars nearly north of it, about one degree distant. With a small glass these may appear as a single star, but a stronger glass will show them separately.

Ursa Major, Ursa Minor, and Canes Venatici

And now for Ursa Major and the Great Dipper, Draco, Cepheus, Cassiopeia, and the other constellations represented on the circular map, being rather too near the horizon for effective observation at this time of the year. First, as the easiest object, look at the star in the middle of the handle of the Dipper (this handle forms the tail of Ursa Major), and a little attention will show you, without the aid of a glass, if your eye-sight is good, that the star appears double. A smaller star seems to be almost in contact with it. The larger of these two stars is called Mizar and the smaller Alcor the Horse and his Rider the Arabs said. Your glass will, of course, greatly increase the distance between Alcor and Mizar, and will also bring out a clear difference of color distinguishing them. They are not true double stars, in that do not revolve about each other, but only appear so by happenstance. Mizar itself is a true double star, resolvable in a small telescope.

Now, if you have a very powerful glass, you may be able to see the Sidus Ludovicianum, a minute star which a German astronomer discovered more than a hundred and fifty years ago, and, strangely enough, taking it for a planet, named it after a German prince. The

45

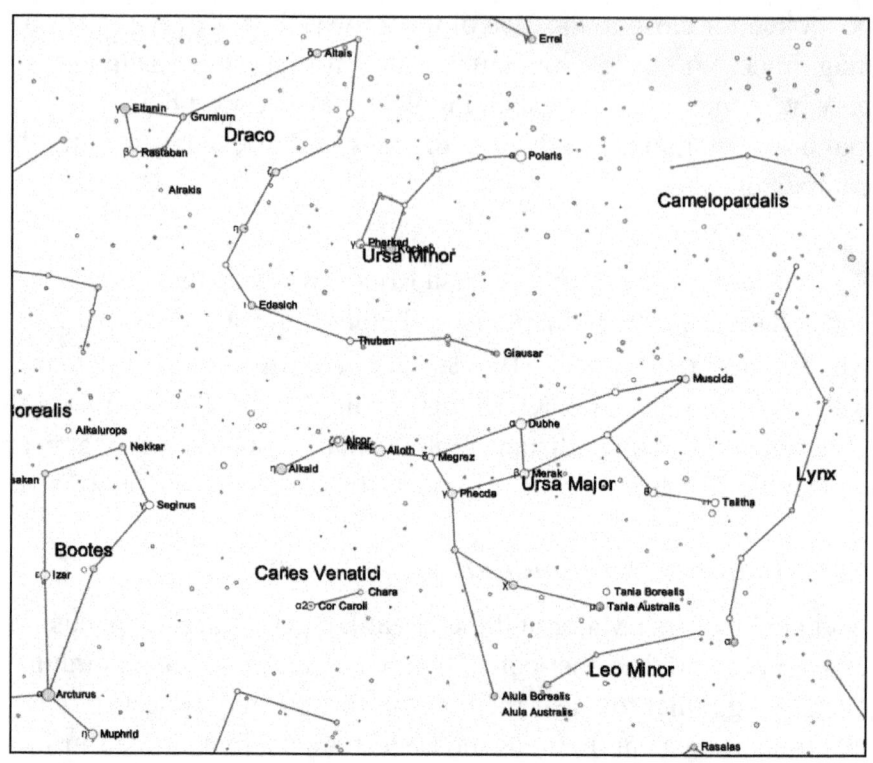

Map 6 - Ursa Major and Minor, Polaris, and Canes Venatici

position of the Sidus Ludovicianum, with reference to Mizar and Alcor, is represented in the accompanying figure. Good binoculars cannot fail to show it.

Sweep along the whole length of the Dipper's handle, and you will discover many fine fields of stars. Then look at the star Alpha (α) Ursa Majoris, or Dubhe in the outer edge of the bowl nearest to the pole-star. There is a faint star, of about the eighth magnitude, near it, in the direction of Beta (β) Ursa Majoris, or Merak. This will prove a very difficult test. You will have to try it with averted vision. Its distance is a little over half that between Mizar and Alcor. It is of a reddish color.

You will notice nearly overhead three pairs of pretty bright stars in a long, bending row, about half-way between Leo and the Dipper. These mark three of Ursa Major's feet, and each of the pairs is well worth looking at with a glass, as they are beautifully grouped with stars invisible to the naked eye. The letters used to designate the stars forming these pairs will be found upon our map of Ursa Major. The scattered group of faint stars beyond the bowl of the Dipper forms the Bear's head, and you will find that also a field worth a few minutes' exploration.

The two bears, Ursa Major and Ursa Minor, swinging around the pole of the heavens, have been conspicuous in the star-lore of all ages. According to fable, they represent

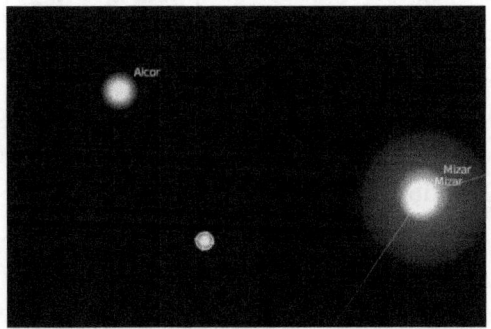

Mizar, Alcor, and the Sidus Ludovicianum.

the nymph Calisto, with whom Jupiter was in love, and her son Areas, who were both turned into bears by Hera (or Juno), whereupon Zeus (Jupiter), being unable to restore their form, did the next best thing he could by placing them among the stars. Ursa Major is Calisto, or Helica, as the Greeks called the constellation. The Greek name of Ursa Minor was Cynosura. The use of the pole-star in navigation dates back at least to the time of the Phoenicians. The observer will note the uncomfortable position of Ursa Minor, attached to the pole by the end of its long tail.

Beneath the handle of the Dipper, you will find the faint constellation Canes Venatici, the hunting dogs of the herdsman Bootes, represented by the principal stars Cor Caroli and Chara. Just northward, towards the Dipper, denoted on our map with a surrounding circle, you may note a star of sixth magnitude with a deeply red hue, made so by the prevalence of carbon compounds in its cool atmosphere. The remarkably beautiful star is aptly named La Superba.

But, after all, no one can expect to derive from astronomy any genuine pleasure or satisfaction unless he is mindful of the real meaning of what he sees. The actual truth seems almost too stupendous for belief. The mind must be brought into an attitude of profound contemplation in order to appreciate it. From this globe we can look out in every direction into the open and boundless universe. Blinded and dazzled during the day by the blaze of that star, of which the earth is a near and humble dependent, we are shut in as by a curtain. But at night, when our own star is hidden, our vision ranges into the depths of creation, and we behold them sparkling with a multitude of other suns.

With so simple an aid as that of binoculars we penetrate still deeper into the profundities of space, and thousands more of these strange, far-away suns come into sight. They are arranged in pairs, sets, rows, streams, and clusters. Here they gleam alone in distant splendor, there they glow and flash in mighty swarms. Here is a celestial city whose temples are suns, and whose streets are the pathways of light.

The Stars of Summer

Let us now suppose that the Earth has advanced for three months in its orbit since we studied the stars of spring, and that, in consequence, the heavens have made one quarter of an apparent revolution. Then we shall find that the stars which in spring shone above the western horizon have been carried down out of sight, while the constellations that were then in the east have now climbed to the zenith, or passed over to the west, and a fresh set of stars has taken their place in the east. In the present chapter we shall deal with what may be called the stars of summer; and, in order to furnish occupation for the observer with binoculars throughout the summer months, I have chosen the constellations to explore such that some shall be favorably situated in each of the months of June, July, and August.

Map 7 represents the heavens at midnight on the 1st of June ; at eleven o'clock, on the 15th of June; at ten o'clock, on the 1st of July; at nine o'clock, on the 15th of July ; and at eight o'clock, on the 1st of August. Remembering that the center of the map is the point over his head, and that the edge of it represents the circle of the horizon, the reader, by a little attention and comparison with the sky, will be able to fix in his mind the relative placement of the various constellations. The maps that follow will show him these constellations on a larger scale, and give him the names of their chief stars and celestial highlights.

The observer need not wait until midnight on the 1st of June in order to find some of the constellations included in our map. Earlier in the evening, at about that date, say at nine o'clock, he will be able to see many of these constellations, but he must look for them farther toward the east than they are represented in the map. The bright stars in Bootes and Virgo, for instance, instead of being over in the southwest, as in the map, will be near the meridian, while Lyra, instead of shining high overhead, will be found climbing up out of the northeast. It would be well to begin at

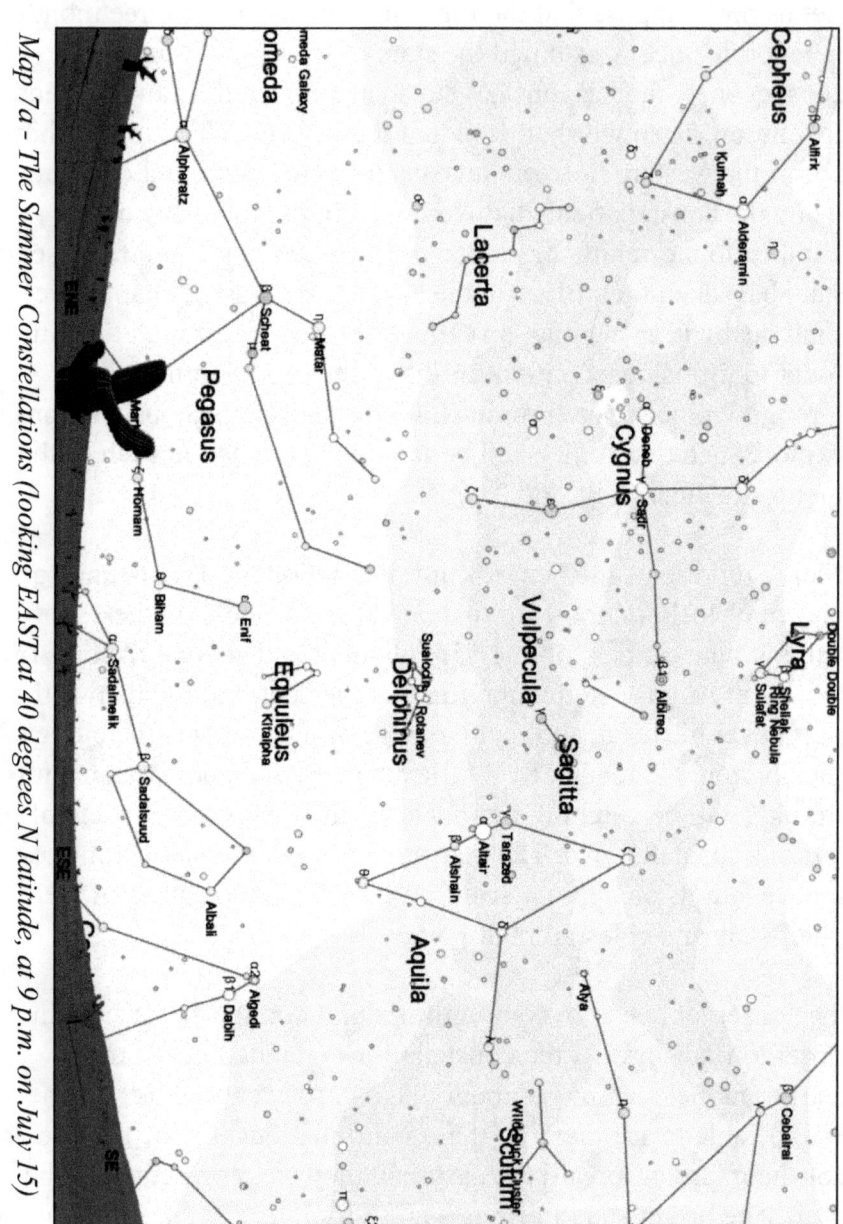

Map 7a - The Summer Constellations (looking EAST at 40 degrees N latitude, at 9 p.m. on July 15)

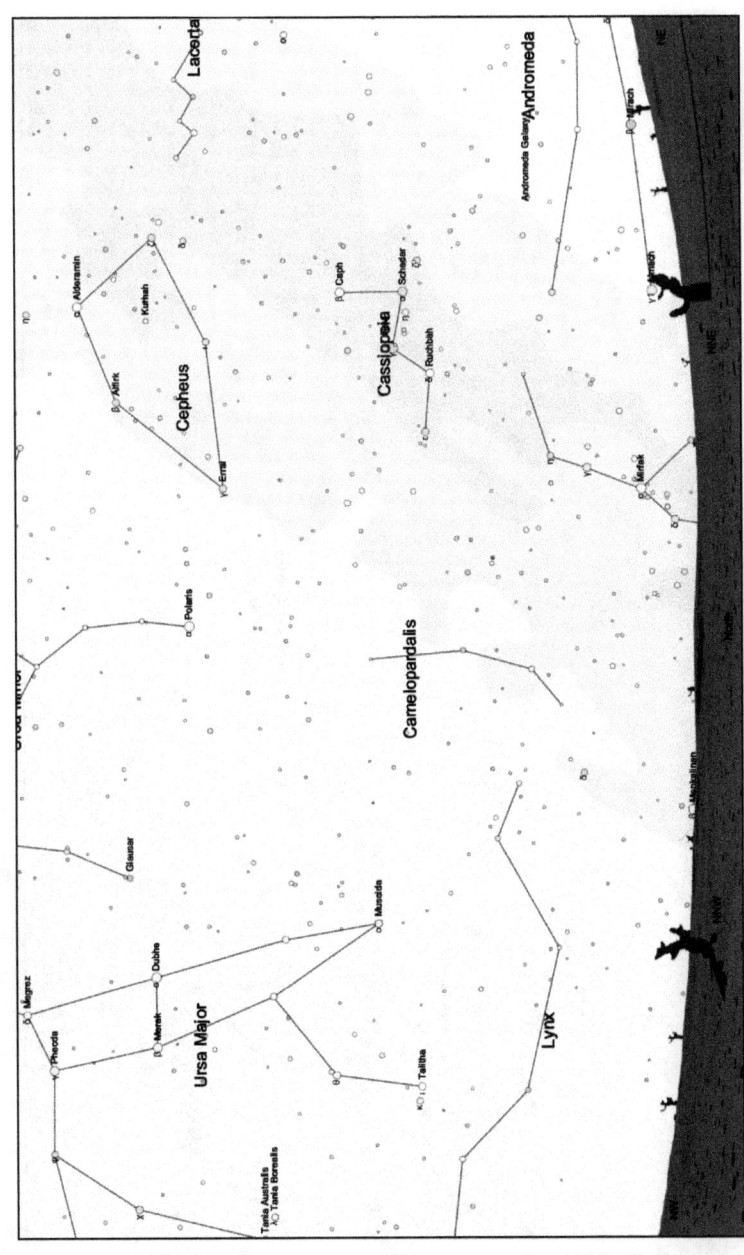

Map 7b - The Summer Constellations (looking NORTH at 40 degrees N latitude, at 9 p.m. on July

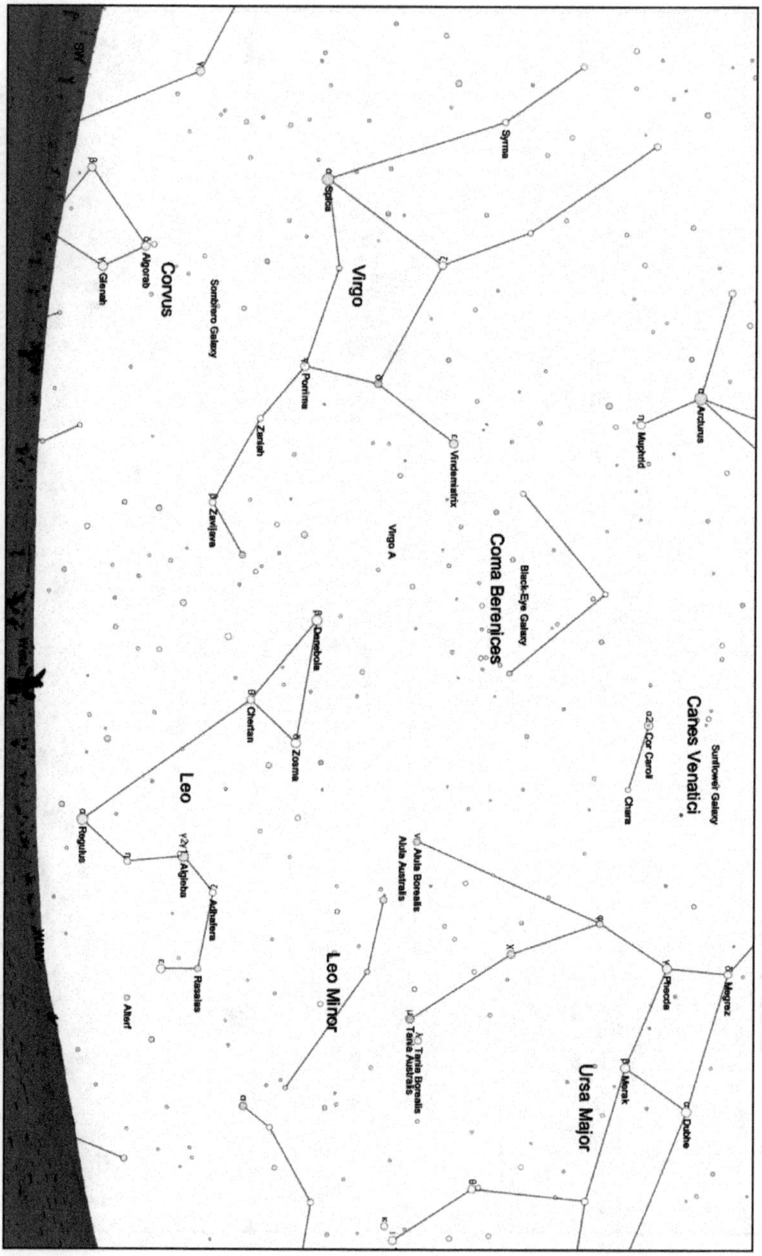

Map 7c - The Summer Constellations (looking WEST at 40 degrees N latitude, at 9 p.m. on July 15)

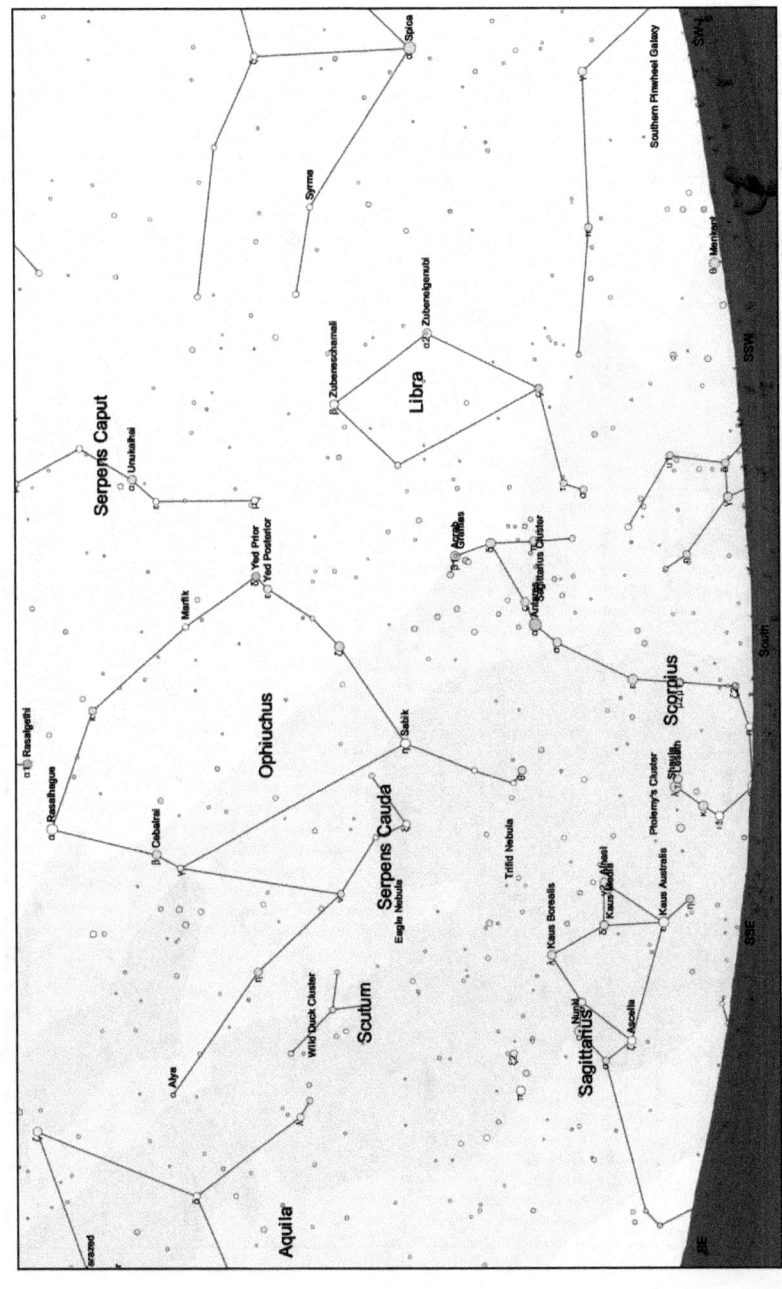

Map 7d - The Summer Constellations (looking SOUTH at 40 degrees N latitude, at 9 p.m. on July

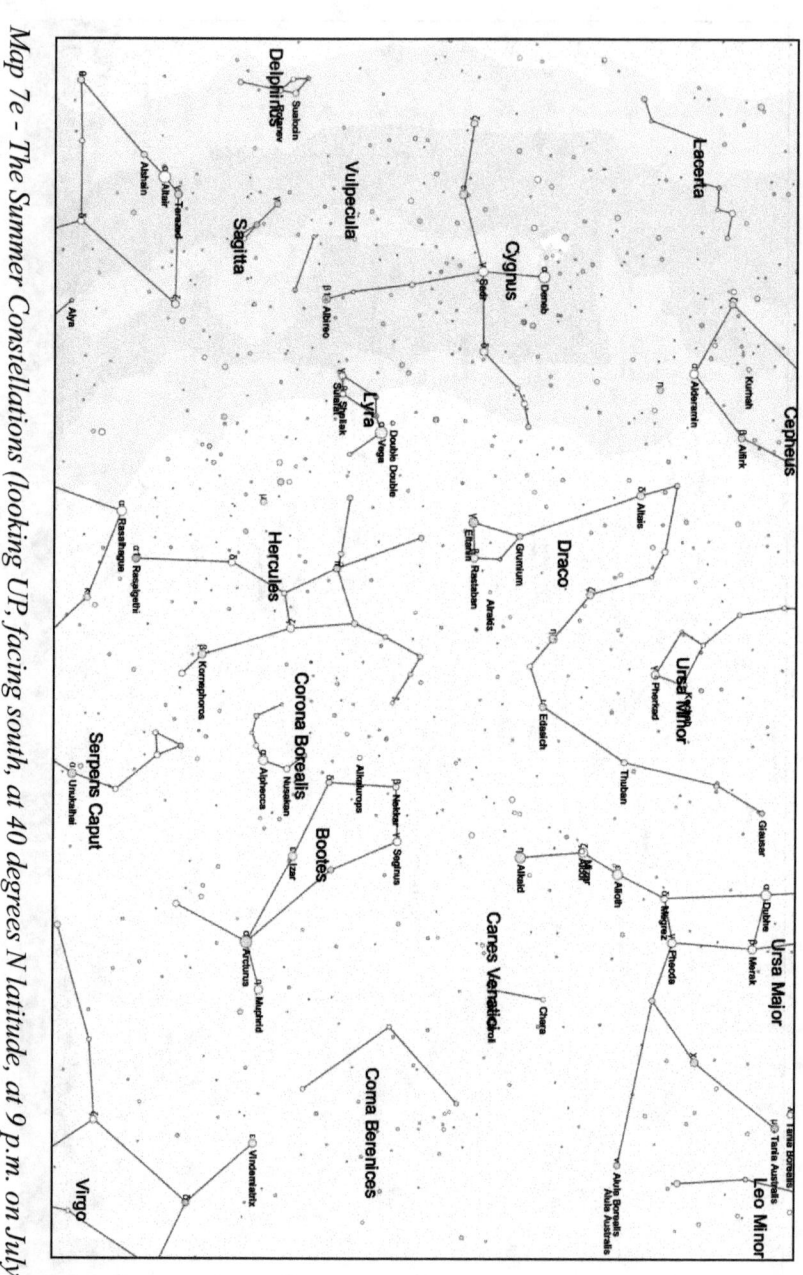

Map 7e - The Summer Constellations (looking UP, facing south, at 40 degrees N latitude, at 9 p.m. on July 15)

54

nine o'clock, about the 1st of June, and watch the motions of the heavens for two or three hours. At the commencement of the observations you will find the stars in Bootes, Virgo, and Lyra in the positions I have just mentioned, while half-way down the western sky will be seen the Sickle of Leo. The brilliant Procyon and Capella will be found almost ready to set in the west and northwest, respectively. Between Procyon and Capella, and higher above the horizon, shine the twin stars in Gemini.

In an hour Procyon, Capella, and the Twins will be setting, and Spica will be well past the meridian. In another hour the observer will perceive that the constellations are approaching the places given to them in our map, and at midnight he will find them all in their assigned positions.

A single evening spent in observations of this sort will teach him more about the places of the stars than he could learn from a dozen books.

Scorpius, the Scorpion

Now find the constellation Scorpius (or Scorpio), and its chief star Antares. The map shows you where to look for it at midnight on the 1st of June. If you prefer to begin at nine o'clock at that date, then, instead of looking directly in the south for Scorpio, you must expect to see it just rising in the southeast. You will recognize Antares by its fiery color, as well as by the striking arrangement of its surrounding stars.

There are few constellations which bear so close a resemblance to the objects they are named after as Scorpius. It does not require much exercise of the imagination to see in this long, winding trail of stars a gigantic scorpion, with its head to the west, and flourishing its upraised sting that glitters with a pair of twin stars, as if ready to strike. Readers of the old story of Phaeton's disastrous attempt to drive the chariot of the Sun for a day will remember it was the sight of this threatening monster that so terrified the ambitious youth as he dashed along the Zodiac, that he lost control of Apollo's horses, and came near burning the earth up by running the Sun into it.

Antares rather gains in redness when viewed with binoculars. Its color is very remarkable, and it is a curious circumstance that with powerful telescopes a small star is seen apparently almost touching it. The spectroscopic appearance of Antares suggest the existence of a powerfully absorptive atmosphere, and which are believed on various grounds to be "in the last visible stage of cooling "; in other words, entering the last stages of its life. This great, red-giant star exceeds our sun in size, and no one can help feeling the sublime nature of those studies which give us reason to think that here we can actually behold the expiring throes of a star. Only, the lifetime of a sun is many billions of years, and its gradual extinction, even after it has reached a stage as advanced as that of Antares is supposed to be, may occupy a longer time than the whole duration of the human race.

A little close inspection with the naked eye will show three fifth- or sixth-magnitude stars above Antares and Alniyat, which form, with those stars, the figure of an irregular pentagon. Binoculars show this figure very plainly. The nearest of these stars to Antares, the one directly above it, is known by the number 22, and belongs to Scorpio, while the farthest away, which marks the northernmost corner of the pentagon, is Rho (ρ) in Ophiuchus. Try your binoculars upon the two stars just named. Take 22 first. You will without much difficulty perceive that it has a little star under its wing, below and to the right, and more than twice as far away above it there is another faint star. Then turn to Rho. Look sharp and you will catch sight of two companion stars, one close to Rho on the right and a little below, and the other still closer and directly above. The latter is quite difficult to be seen distinctly, but the sight is a very pretty one.

Your binoculars will show a number of faint stars scattered around Antares. Turn now to Beta (β) Scorpii, also called Achrab or Graffias. A very pretty pair of stars will be seen hanging below Beta. Sweeping downward from this point to the horizon you will find many beautiful star-fields. The star marked Nu (ν) is a double which you will be able to separate with binoculars, the distance between its components being 40 arcseconds (one arcsecond is 1/3600 of a degree).

And next let us look at a star cluster. You will see on Map No. 8 an object marked M4, near Antares. Its designation means that it is No. 4 in Messier's catalogue of nebulae. It is not a nebula, but a closely compacted cluster of stars. With binoculars, if you are looking in a clear and moonless night, you will see it as a curious nebulous speck. You may see it blaze brighter toward the center. It is, in fact, a globular star cluster in which thousands of suns are associated together into splendid assemblages. The object above and to the right of Antares, marked in the map as M80., is a also a globular cluster, although is will appear in binoculars as a mere wisp of light. Yet there is a pretty array of small stars in its neighborhood worth looking at. Besides, this cluster is of special interest, because in 1860 a star suddenly took its place. At least, that is what seemed to have happened. What really did occur, probably, was that a variable or temporary star, called a nova, and ordinarily too faint to be perceived, blazed up as to shine as brightly as the entire cluster. If this star should make its appearance again, it could easily be seen with binoculars.

The quarter of the heavens with which we are now dealing is famous for these celestial conflagrations, if so they may be called. The first temporary star of which there is any record appeared in the constellation of the Scorpion, near the head, 134 B.C. It must have been a most extraordinary phenomenon, for it attracted attention all over the., world, and both Greek and Chinese annals contain descriptions of it. In 393 A.D. a temporary star shone out in the tail of Scorpio. In 827 A.D. Arabian astronomers, under the Caliph Al-Mamoun, the son of Haroun-al-Raschid, who broke into the great pyramid, observed a temporary star, that shone for four months in the constellation of the Scorpion. In 1203 there was a temporary star, of a bluish color, in the tail of Scorpio, and in 1578 another in the head of the constellation. Besides these there are records of the appearance of four temporary stars in the neighboring constellation of Ophiuchus, one of which, that of 1604, is very famous, and will be described later on. It is conceivable that these strange outbursts in and near Scorpio may have had some effect in causing this constellation to be regarded by the ancients as malign in its influence.

Let us follow the bending row of stars from Antares toward the south and east. When you reach the star Mu (μ) you are likely to stop with an exclamation of admiration, for the glass will separate it into two stars that, shining side by side, seem trying to rival each other in brightness. But the next star below marked Zeta (ζ), is even more beautiful. It also separates into two stars, one being reddish and the other bluish in color. The contrast in a clear night is very pleasing. But this is not all. Above the two stars you will notice a curious nebulous speck. Now, if you have powerful binoculars, here is an opportunity to view one of the prettiest sights in the heavens. The glass not only makes the two stars appear brighter, and their colors more pronounced, but it shows a third, fainter star below them, making a small triangle, and brings other still fainter stars into sight, while the nebulous speck above turns into a charmingly beautiful little star-cluster, whose components are so close that their rays are inextricably mingled in a maze of light.

Following the bend of the Scorpion's tail upward, we come to the pair of stars in the stinger. These, of course, are separated well by binoculars. Then let us sweep off to the eastward a little way and find the cluster known as M7. You will see it marked on the map. Above it, and near enough to be included in the same field of view, is M6, a smaller cluster. Both of these have a sparkling appearance with binoculars, and by close attention some of the separate stars in M7 may be detected. These clusters become much more striking and starry looking at higher power, and the curious radiated structure of M7 comes out.

In looking at such objects we can not too often recall to our minds the significance of what we see that these glimmering specks are the lights in the windows of the universe which carry to us, across inconceivable tracts of space, the assurance that we are not alone in the heavens; that all around us, and even on the very confines of immensity, Nature is busy, as she is here, and the laws of light, heat, gravitation (and perhaps of life?), are in full activity.

The Wonders of Sagittarius

The clusters we have just been looking at lie on the borders of Scorpio and Sagittarius. Let us cross over into the latter constellation, which commemorates the centaur Chiron. We are now in another, and even a richer, region of wonders. The Milky-Way, streaming down out of the north-east, pours, in a luminous flood, through Sagittarius, inundating that whole region of the heavens with seeming deeps and shallows, and finally bursting the barriers of the horizon disappears, only to glow with more splendor in the southern hemisphere. The stars *Ascella* (Zeta, ζ), Tau (τ), *Nunki* (Sigma, σ), Phi (φ), *Kaus Australis* Epsilon (ε), *Kaus Media* Delta (δ), *Kaus Borealis* Lambda (λ), and *Alnasl* Gamma (γ), indicate the outlines of a figure sometimes called, the "Tea Pot", which is very evident when the eye has once recognized it. Above the top of the Teapot lies another star Mu (μ). In the region around Mu lie some of the most interesting objects in the sky.

Let us start at Mu (μ). Sweep downward and to the right a little way, and you will be startled by a most singular phenomenon that

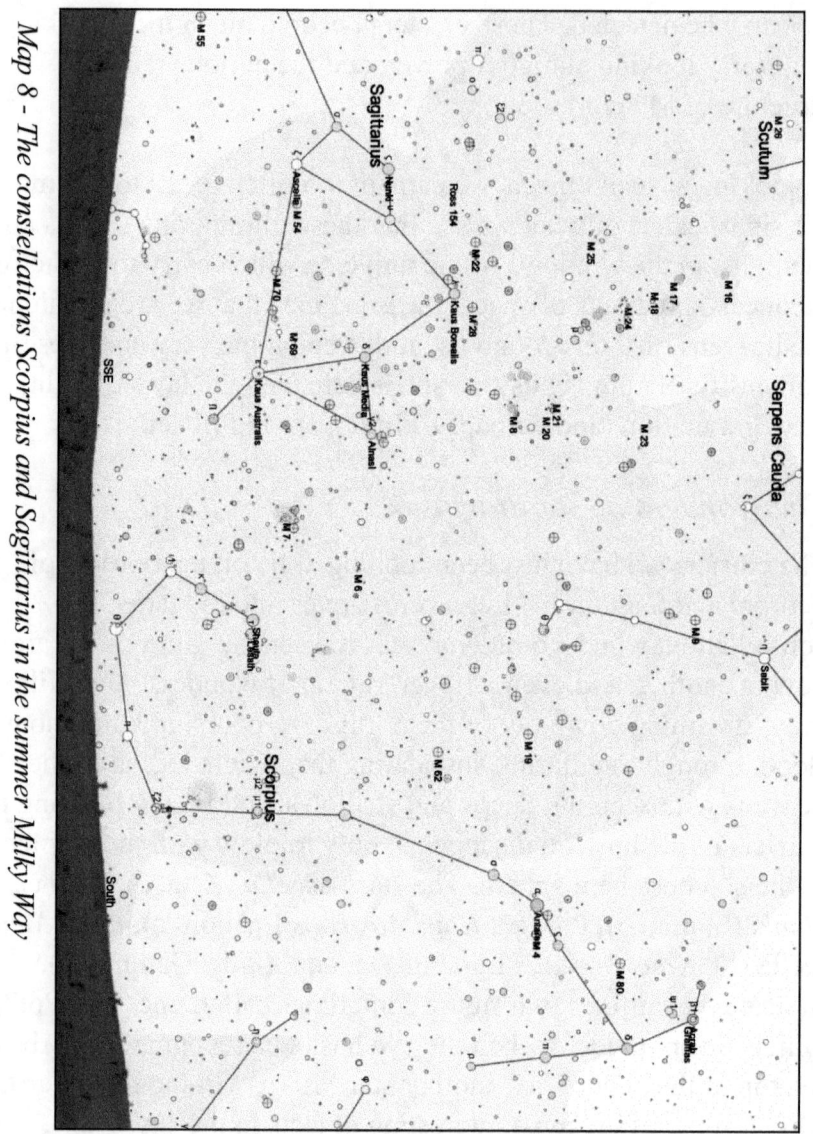

Map 8 - The constellations Scorpius and Sagittarius in the summer Milky Way

has suddenly made its appearance in the field of view of your glass. You may, perhaps, be tempted to congratulate yourself on having got ahead of all the astronomers, and discovered a comet. It is really a combination of a star cluster with a nebula, and is known as M8, the "Lagoon Nebula". Sir John Herschel has described the "nebulous folds and masses" and dark oval gaps which he saw in this nebula with his large telescope at the Cape of Good Hope. But no telescope is needed to make it appear a wonderful object; binoculars reveal much of its marvelous structure.

The reader will recall that we found the summer solstice close to a wonderful star-swarm in the feet of Gemini. Strangely enough the winter solstice is also near a star cluster. It is to be found near a line drawn from M8 to the star Mu (μ), and about one third of the way from the cluster to the star. There is another less conspicuous star-cluster still closer to the solstitial point here, for this part of the heavens teems with rich fields of stars.

On the opposite side of the star Mu, that is to say, above and a little to the left is an entirely different but almost equally attractive spectacle, the Sagittarius Star Cloud M24. Here, again, simple binoculars bring out the innumerable points of light of which the cluster is composed. Do not fail to gaze long and steadily at this island of stars, for much of its beauty becomes evident only after the eye has accustomed itself to disentangle the glimmering rays with which the whole field of view is filled.

Just to the left and a little northwards from *Kaus Borealis,* at the tip of the Tea Pot, look for a tiny hazy patch of stars, which was labeled by Messier as M22.

For these objects, try the method of averted vision, and hundreds of the faint stars will seem to spring into view out of the depths of the sky. The necessity of a perfectly clear night, and the absence of moonlight, cannot be overemphasized for observations such as these. Everybody knows how the moonlight blots out the smaller stars. A slight haziness or smoke in the air produces a similar effect. It is as important to the observer with binoculars to have a transparent atmosphere as it is to one who would use a telescope; but, fortunately, the work of the former is not so much interfered with by currents of air. Always avoid the neighborhood of any bright light. Electric lights in particular are an abomination to stargazers.

The cloud of stars we have just been looking at is in a very rich region of the Milky-Way, in the little group called "Sobieski's Shield", which we have not named upon our map. Sweeping slowly upward from Mu (μ), a little way with the binoculars, we will pass in succession over three nebulous-looking spots. The second of these, counting upward, is the famous Horseshoe nebula (M17). Its wonders are beyond the reach of our instrument, but its place may be recognized. Look carefully all around this region, and you will perceive that the old gods, who traveled this road (the Milky Way was sometimes called the pathway of the gods), trod upon golden sands. Off a little way to the east you will find the rich cluster M25. But do not imagine the thousands of stars that your binoculars reveal comprise all the riches of this region of the heavens. You can ply the powers of the greatest telescope and still not exhaust its wealth.

The milky look of the background of the Galaxy is, of course, caused by the intermingled radiations of inconceivably numerous stars, thousands of which become separately visible, the number thus distinguishable varying with the size of the instrument. Binoculars or a small telescope cannot sound these starry deeps to the bottom.

The groups of stars forming the eastern half of the constellation of Sagittarius are worth sweeping over with the binoculars, as a number of pretty pairs may be found there. Sagittarius stands in the old star-maps as a centaur, half-horse-half-man, facing the west, with bow drawn, and arrow pointed at the Scorpion.

Ophiuchus and Serpens

Next let's pass to the constellations adjoining Scorpio and Sagittarius on the north, Ophiuchus and Serpens, the Serpent. These constellations, as our map shows, are curiously intermixed. Serpens is unique among constellations in that it's split in two. The imagination of the old star-gazers, who named them, saw here the figure of a giant grasping a writhing serpent with his hands. The head of the serpent is

under the Northern Crown, and is called Serpens Caput. Its tail ends over the star-gemmed region that we have just described, called "Sobieski's Shield", and is called Serpens Cauda. Ophiuchus stands, as figured in some star atlases, upon the back of the Scorpion, holding the serpent with one hand below the neck, and with the other near the tail. The giant's face is toward the observer, and the star Alpha (α), also called Ras Alhague, shines in his forehead, while Cebalrai, or Beta, (β) and Gamma (γ) mark his right shoulder. Ophiuchus has been held to represent the famous physician Aesculapius. One may well repress the tendency to smile at these fanciful legends when he reflects upon their antiquity. There is no doubt that this double constellation is at least three thousand years old, which means for thirty centuries the imagination of men has continued to shape these stars into the figures of a gigantic man struggling with a huge serpent. Like many other of the constellations it has proved longer-lived than the mightiest nations. While Greece flourished and decayed, while Rome rose and fell, while the scepter of civilization has passed from race to race, these starry creations of fancy have shone on unchanged. The mind that would ignore them now deserves compassion.

Just south of the star *Sabik* in Ophiuchus lies the spot where one of the most famous temporary stars on record appeared in the year

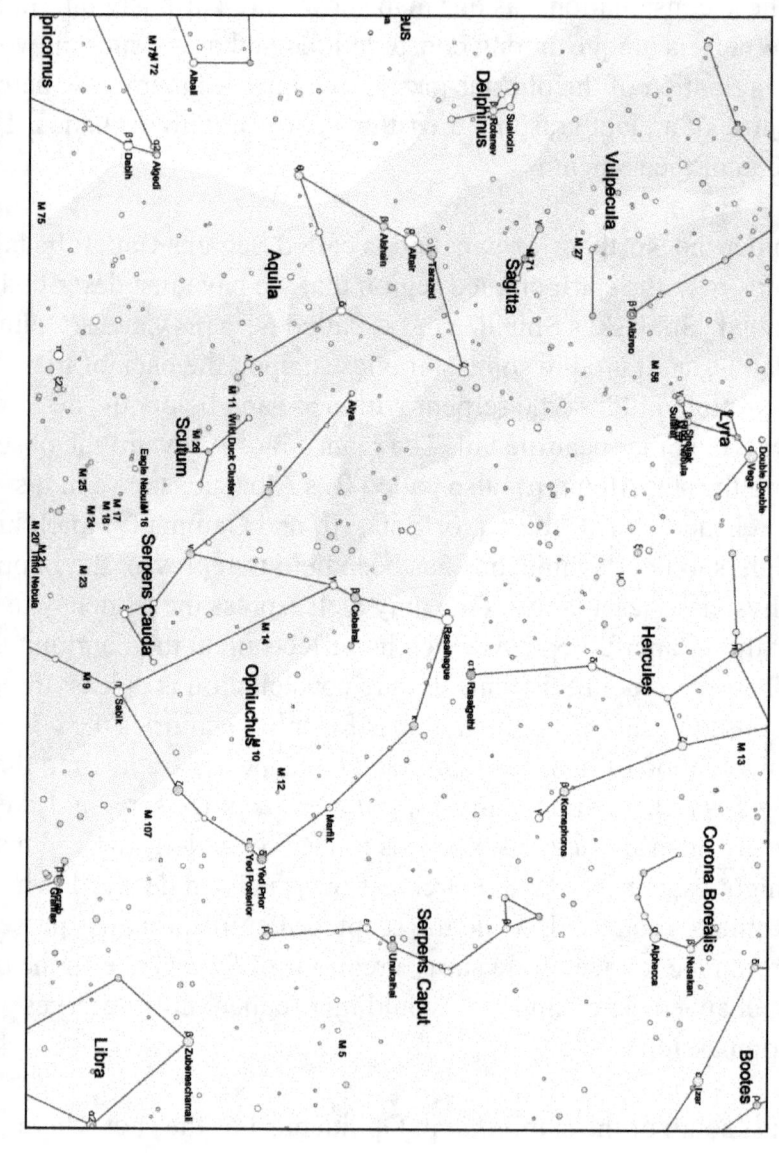

Map 9 – The constellations Ophiuchus and Serpens (Caput, the head, and Cauda, the tail).

1604. At first it was far brighter than any other star in the heavens, but it quickly faded, and in a little over a year disappeared. It is particularly interesting, because the great astronomer and mathematician Kepler wrote a curious book about it. Some of the philosophers of the day argued that the sudden outburst of the wonderful star was caused by the chance meeting of atoms. Kepler vigorously disagreed, and his reply was characteristic, as well as amusing:

"I will tell those disputants, my opponents, not my own opinion, but my wife's. Yesterday, when I was weary with writing, my mind being quite dusty with considering these atoms, I was called to supper, and a salad I had asked for was set before me. 'It seems, then', said I, aloud, 'that if pewter dishes, leaves of lettuce, grains of salt, drops of water, vinegar and oil, and slices of egg, had been flying about in the air from all eternity, it might at last happen by chance that there would come a salad. 'Yes,' says my wife, 'but not so nice and well-dressed as this of mine is.' "

Investigation by modern astronomers into the nature of stars revealed that Kepler's star was one of the most violent events in all the heavens: a supernova, the dying remnant of a massive star which has finally run out of fuel to hold itself up from the relentless force of gravity.

While there are few objects of special interest for the observer with binoculars in Ophiuchus, he will find it worth while to sweep over it for what he may pick up, and, in particular, he should look at the group of stars southeast of Beta (β) and Gamma (γ). These stars have been shaped into a little modern asterism called *Taurus Poniatowskii*, and it will be noticed that five of them mark the outlines of a letter "V", resembling a bull's head, much like the Hyades in the constellation Taurus.

Also look at the stars in the head of Serpens, several of which form a figure like a letter X. A little west of Eta (η), in the tail of

Serpens, is a beautiful swarm of little stars, called M11, the "Wild Duck" cluster in the neighboring constellation Scutum.

Do not fail to notice the remarkable subdivisions of the Milky-Way in this neighborhood. Its current seems divided into numerous channels and bays, interspersed with gaps that might be likened to islands. This complicated structure of the Milky-Way extends downward to the horizon, and upward through the constellation Cygnus, and of its phenomenal appearance in that region we shall have more to say further on.

Hercules, and the Great Cluster

Directly north of Ophiuchus is the constellation Hercules, interesting as occupying that part of the heavens toward which the proper motion of the sun is bearing the earth and its fellow planets, at the rate of 16.5 kilometers per second in a stupendous voyage through space.

In the accompanying Map 10 we have represented the beautiful constellations Lyra and the Northern Crown, lying on either side of Hercules. The bottom of the map is toward the south, the right-hand side is west, and the left-hand side east. It is important to keep these directions in mind, in comparing the map with the sky. For instance, the observer must not expect to look into the south and see Hercules half-way up the sky, with Lyra a little east of it; he must look for Hercules nearly overhead, and Lyra a little east of the zenith. The same precautions are not necessary in using the maps of Scorpio, Sagittarius, and Ophiuchus, because those constellations are nearer the horizon, and so the observer does not have to imagine the map as being suspended over his head.

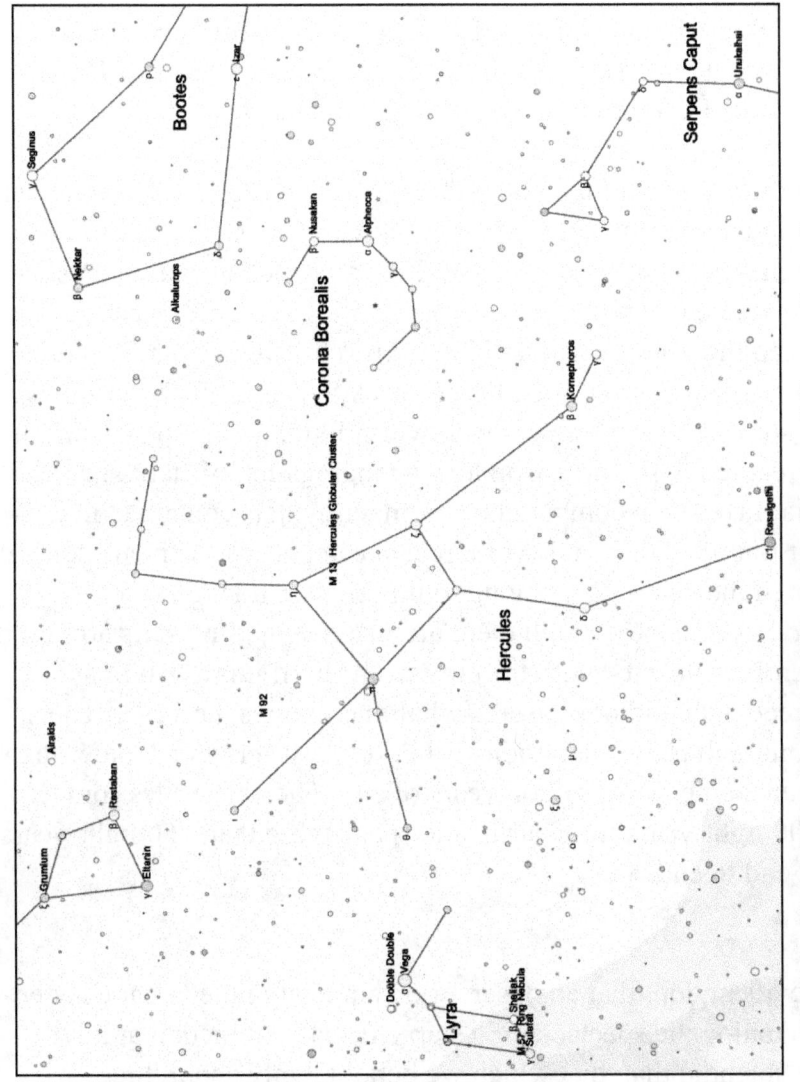

Map 10 - The constellations Hercules, Lyra, and Corona Borealis.

The name Hercules indicates the mythological origin of the constellation, and yet the Greeks did not know it by that name, for the poet Aratus calls it "the Phantom whose name none can tell." Corona Borealis, the "Northern Crown", according to fable, was the celebrated crown of Ariadne, and Lyra was the harp of Orpheus himself, with whose sweet music he charmed the hosts of Hades, and persuaded Pluto to yield up to him his lost love Eurydice.

With the aid of the map you will be able to recognize the principal stars and star-groups in Hercules, and will find many interesting combinations of stars for yourself. An object of special interest is the celebrated star-cluster M13. You will find it on the map between the stars Eta (η) and Zeta (ζ). While binoculars will only show it as a nebulous and minute speck, lying nearly between two little stars, it is nevertheless well worth looking for, on account of the great renown of this wonderful congregation of stars. Sir William Herschel computed the number of stars contained in it as about fourteen thousand. More recent calculations estimate several hundred thousand stars belong to this cluster. It is roughly spherical in shape, though there are many straggling stars around it evidently connected with the cluster. In short, it is a ball of suns. The reader should not mistake what that implies, however. These suns, though truly solar bodies, are probably very much older than our sun, some 10-12 billion years of age. The cluster lies some 25,000 light-years from earth, and spans more than 150 light-years from end to end.

It is evident, too, that an observer on a planet in the Great Cluster would enjoy the spectacle of a starry firmament incomparably more splendid than that which we behold. Only about three thousand stars are visible to our unaided eyes at once on any clear night, and of those only a few are conspicuous, and two thirds are so faint that they require some attention in order to be distinguished. But the spectator at the center of the Hercules cluster would behold some ten thousand stars at once, the faintest of which would be five times as brilliant as the brightest star in our sky, while the brighter ones would blaze like nearing suns. One effect of this flood of starlight may be to shut out from our observer's eyes all the stars of the outside universe. They would be effaced in the blaze of his sky, and he would be, in a manner, shut up within his own little star-system, knowing nothing of the greater universe beyond, in which we behold the blazing stars in his sky, diminished by distance into a faint speck, floating like a silvery mote in a sunbeam.

The Coma Berenices Star Cluster in binoculars

If our observer's planet, instead of being situated in the center of the cluster, circled around one of the stars at the outer edge of it, the appearance of his sky would be, in some respects, still more wonderful. Less than half of his sky would be filled, at any time, by the stars of the cluster, the other half opening upon outer space and appearing by comparison almost starless a vast, cavernous expanse, with a few faint glimmerings out of its gloomy depths. The plane of the orbit of his planet being supposed to pass through the center of the spherical system, our observer would, during his year, behold the night at one season blazing with the splendors of the clustered suns, and at another empty and faintly lighted with the soft glow of the Milky-Way and the feeble flickering of distant stars, scattered over the dark sky. The position of the orbit, and the inclination of the planet's axis might be such that the glories of the cluster would not be visible from one of its hemispheres, necessitating a journey to the other side of the globe to behold them.

Larger binoculars will give you a more satisfactory view of M13, and there can be no possibility of mistaking it for a star. Compare this compact cluster, which only a 6 or 8-inch telescope can partially resolve into hundreds of component stars, with M7 and M24, described before, in order to comprehend the wide variety in the structure of these aggregations of stars.

Lyra, the Lyre

The Northern Crown, although a strikingly beautiful constellation to the naked eye, offers few attractions to binoculars. Let us turn, then, to Lyra. I have never been able to make up my mind which of three great stars is entitled to precedence Vega, the leading brilliant of Lyra, Arcturus in Bootes, or Capella in Auriga. They are the three leaders of the northern firmament, but which of them should be called the chief, is very hard to say. At any rate, Vega would probably be generally regarded as the most beautiful, on account of the delicate bluish tinge in its light, especially when viewed with a glass. There is no possibility of mistaking this star because of its surpassing brilliance. Two faint stars close to Vega on the east make a beautiful little triangle with it, and thus form a further means of recognition, if any were needed. Your optics will show that the background sky is powdered with stars, fine as the dust of a diamond, all around the neighborhood of Vega, and the longer you gaze the more of these diminutive stars you will discover.

Now direct your glass to the northernmost of the two little stars near Vega, the one marked Epsilon (ε) in the map. You will perceive that it is composed of two stars of almost equal magnitude. If you had a telescope of considerable power, you would find that each of these stars is in turn double. In other words, this wonderful star which appears single to the unassisted eye and double in binoculars, is in reality quadruple, and the four stars composing it are connected in pairs, the members of each pair revolving around their common center while the two pairs in turn circle around a center common to all. With your binoculars, you will be able to see that the other star near Vega, Zeta (ζ), is also double, the distance between its components being three quarters of an arc-minute, while the two stars in Epsilon are a little less than 3.5 arc-minutes apart. The star Beta (β) is remarkably variable in brightness. You may watch these variations, which run through a regular period of nearly 13 days, for yourself. Between Beta and Gamma (γ) lies the beautiful Ring nebula, but it is visible only in 80 mm or larger binoculars, and even then, only as a tenuous speck. A telescope is required to glimpse its full glory.

Let us turn next to the stars in the west. In consulting the accompanying Map 11 of Virgo and Bootes, the observer is supposed to face the southwest, at the hours and dates mentioned above as those to which the circular map corresponds. He will then see the bright star Spica in Virgo not far above the horizon, while Arcturus will be half-way up the sky, and the Northern Crown will be near the zenith.

Virgo

The constellation Virgo is an interesting one in mythological story. Aratus tells us that the Virgin's home was once on earth, where she bore the name of Justice, and in the golden age all men obeyed her. In the silver age her visits to men became less frequent, "no longer finding the spirits of former days"; and, finally, when the bronze age came with the clangor of war:

"Justice, loathing that race of men,
Winged her flight to heaven ; and fixed
Her station in that region
Where still by night is seen
The Virgin goddess near to bright Bootes."

The chief star of Virgo, Spica, is remarkable for its pure white light. To my eye there is no conspicuous star in the sky equal to it in this respect, and it gains in beauty when viewed with a glass. With the aid of the map the reader will find the celebrated binary star Gamma (γ) Virginis, Porrima, although he will not be able to separate its components without a telescope. It is a curious fact that the star Epsilon (ε) in Virgo, called Vindemiatrix, has for many ages been known as the Grape-Gatherer. It has borne this name in Greek, in Latin, in Persian, and in Arabic, the origin of the appellation undoubtedly being that it was observed to rise just before the sun in the season of the grape harvesting.

The stars Vindemiatrix, δ (Delta), Porrima, Zaniah, and Zavijava, mark two sides of a quadrilateral figure of which the opposite corner is indicated by Denebola in the tail of Leo. Within this quadrilateral lies the marvelous field of the galaxies known as the Virgo Cluster, a region where with a modest telescope one may find dozens of spiral and elliptical galaxies thronging together, each hosting hundreds of billions of stars. Unfortunately, these galaxies are beyond the reach of most small binoculars, but it is worth while to know where this curious region is, even if we can not behold the wonders it contains. The stars Omicron (o) and Pi (π) form a little group that mark the head of Virgo.

The autumnal equinox, or the place where the sun crosses the equator of the heavens on his southerly journey about the 21st of September, is situated nearly between the stars Zaniah and

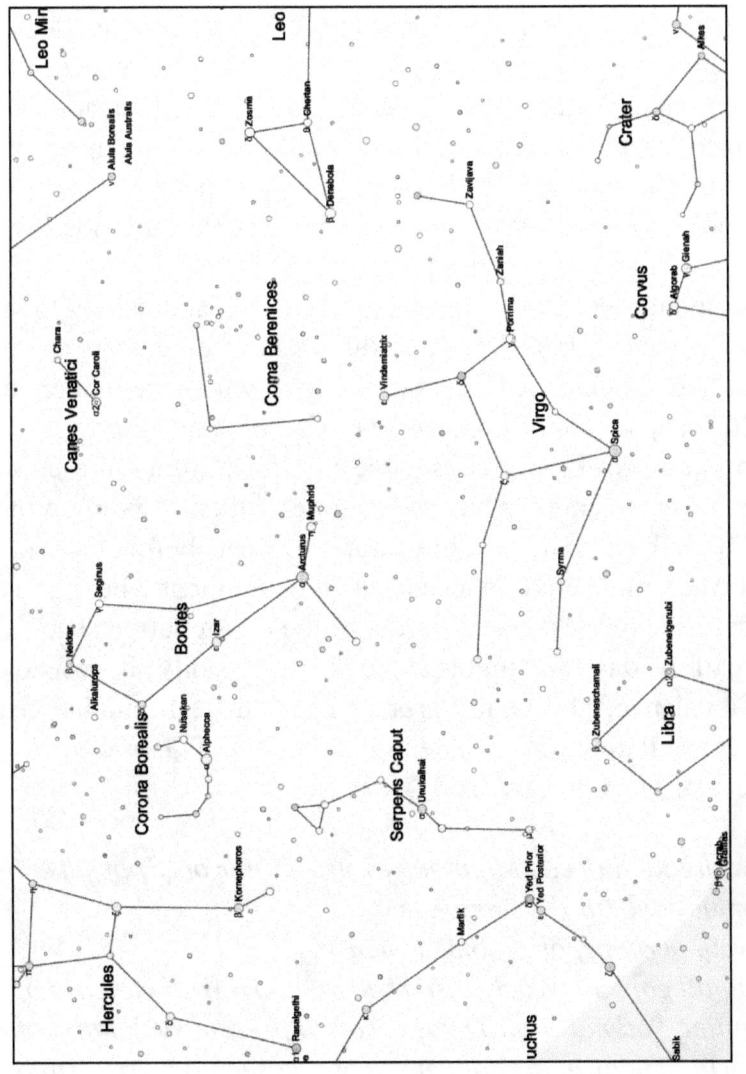

Map 11 - The constellations Bootes, Virgo, and Libra.

73

Zavijava, a little below the line joining them, and somewhat nearer to Zavijava. Both stars are almost exactly upon the equator of the heavens. The celestial equator lies directly above the Earth's equator.

The constellation Libra, lying between Virgo and Scorpio, does not contain much to attract our attention. Its two chief stars, beta (Zubeneschamali) and alpha (Zubenelgenubi), may be readily recognized west of and above the head of Scorpio. The upper one of the two, Zubeneschamali, has a singular greenish tint, while the lower one, Zubenelgenubi, is a very pretty double for binoculars.

The constellation of Libra appears to have been added later than the other eleven members of the zodiac. Its two chief stars at one time marked the extended claws of Scorpio, which were afterward cut off (perhaps the monster proved too horrible even for its inventors) to form Libra. As its name signifies, Libra represents a balance, or scale, and this fact seems to refer the invention of the constellation back to at least three hundred years before Christ, when the autumnal equinox occurred at the moment when the sun was just crossing the western border of the constellation. The equality of the days and nights at that season readily suggests the idea of a balance. Milton, in *Paradise Lost*, suggests another origin for the constellation of the scale in the account of Gabriel's discovery of Satan in paradise:

"... Now dreadful deeds Might have ensued, nor only paradise In this commotion, but the starry cope
Of heaven, perhaps, or all the elements
At least had gone to wrack, disturbed and torn With violence of this conflict, had not soon The Eternal, to prevent such horrid fray, Hung forth in heaven his golden scales, yet seen Betwixt Astrea and the Scorpion sign."

Coma Berenices, and Bootes, The Herdsman

Just north of Virgo's head, you will see the glimmering of Coma Berenice's (Berenice's Hair). This little constellation was included among those described in the chapter on "The Stars of Spring," but it is worth looking at again in the early summer, on moonless nights, when the singular arrangement of the brighter members of the cluster at once strikes the eye.

Bootes, whose brilliant star, Arcturus, occupies the center of our map, also possesses a curious mythical history. It is called by the Greeks the Bear-Driver, because it seems continually to chase Ursa Major, the Great Bear, in his path around the pole. The story is that Bootes was the son of the nymph Calisto, whom Hera (or Juno in the Roman

pantheon), in one of her customary fits of jealousy, turned into a bear. Bootes, who had become a famous hunter, one day roused a bear from her lair, and, not knowing that it was his mother, was about to kill her, when Zeus (Jupiter) came to the rescue and snatched them both up into the sky, where they have shone ever since. Lucan refers to this story when, describing Brutus's visit to Cato at night, he fixes the time by the position of these constellations in the heavens:

> " 'Twas when the solemn dead of night came on,
> When bright Calisto, with her shining son,
> Now half the circle round the pole had run. "

Bootes is not specially interesting for our purposes, except for the splendor of Arcturus. This star has possessed a peculiar charm for me ever since boyhood, when, having read a description of it, I felt an eager desire to see it. As my search for it chanced to begin at a season when Arcturus did not rise till after a boy's bed-time, I was for a long time disappointed, and I shall never forget the start of surprise and almost of awe with which I finally caught sight of it, one spring evening, shooting its flaming rays through the boughs of an apple-orchard, like a star on fire. The light from Arcturus was used, when magnified upon an electric cell, to trigger the lighting of the 1933 Chicago World's Fair.

When near the horizon, Arcturus has a remarkably reddish color; but, after it has attained a high elevation in the sky, it appears rather a deep yellow than red. There is a scattered cluster of small stars surrounding Arcturus, forming an admirable spectacle with binoculars on a clear night. To see these stars well, the glass should be slowly moved about. Many of them are hidden by the glare of Arcturus. The little group of stars near the end of the handle of the Great Dipper, or, what is the same thing, the tail of the Great Bear, marks the upraised hand of Bootes. Between Berenice's Hair and the tail of the Bear you will see a small constellation called Canes Venatici, the Hunting-Dogs. On the old star-maps Bootes is represented as holding these dogs with a leash, while they are straining in chase of the Bear. You will find some pretty groupings of stars in this constellation, including the beautiful poppy-red star *La Superba*, which we met in our tour of the spring sky.

Cygnus (The Northern Cross); Aquila

And now we will turn to the east. Map 12 shows Cygnus (a constellation especially remarkable for the large and striking figure that it contains, called the Northern Cross), Aquila the Eagle, Delphinus, the Dolphin, and the tiny constellations Sagitta and Vulpecula. In consulting the map, the observer is supposed to face toward the east. In Aquila the curious arrangement of two stars on either side of the chief star of the constellation, called Altair, at once attracts the eye. Within a circle including the two attendants of Altair you will probably be able to see with the naked eye only two or three stars in addition to the three large ones. Now turn your glass upon the same spot, and you will see eight or ten times as many stars. Watch the star marked Eta (η), and you will find that its light is variable, being sometimes more than twice as bright as at other times. Its changes are periodical, and occupy a little over a week.

The Eagle is fabled to have been the bird that Jupiter kept beside his throne. A constellation called Antinous, invented by Tycho Brahe, is represented on some maps as occupying the lower portion of the space given to Aquila.

The Dolphin is an interesting little constellation, and the ancients said it represented the very animal on whose back the famous musician Arion rode through the sea after his escape from the sailors who tried to murder him. But some modern has dubbed it with the less romantic name of Job's Coffin, by which it is sometimes called. It presents a very pretty sight to binoculars.

Cygnus, the swan, is a constellation whose mythological history is not specially interesting, although, as remarked above, it contains one of the most clearly marked figures to be found among the stars, the famous Northern Cross. The outlines of this cross are marked with great distinctness by the stars Deneb, Gienah, Sadr, Delta (δ), and Albireo. Albireo, is one of the most beautiful double stars in the heavens. The components are sharply contrasted in color, the larger star being golden-yellow, while the smaller one is a deep, rich blue. With binoculars of 50 mm aperture and magnifying 10x I have sometimes been able to divide this pair, and to recognize the blue color of the smaller star. It will be found a severe test for such a glass.

About half-way from Albireo to the star Zeta (ζ) in Aquila is a very curious little group, called Brocchi's Cluster. Also called the Coathanger, it consists of six or seven stars in a straight row, with a garland of other stars hanging from the center, just like its informal namesake. It is a marvelous sight in binoculars.

I have indicated the place of the celebrated star 61 Cygni in the map, because of the interest attaching to it as one of the nearest to

77

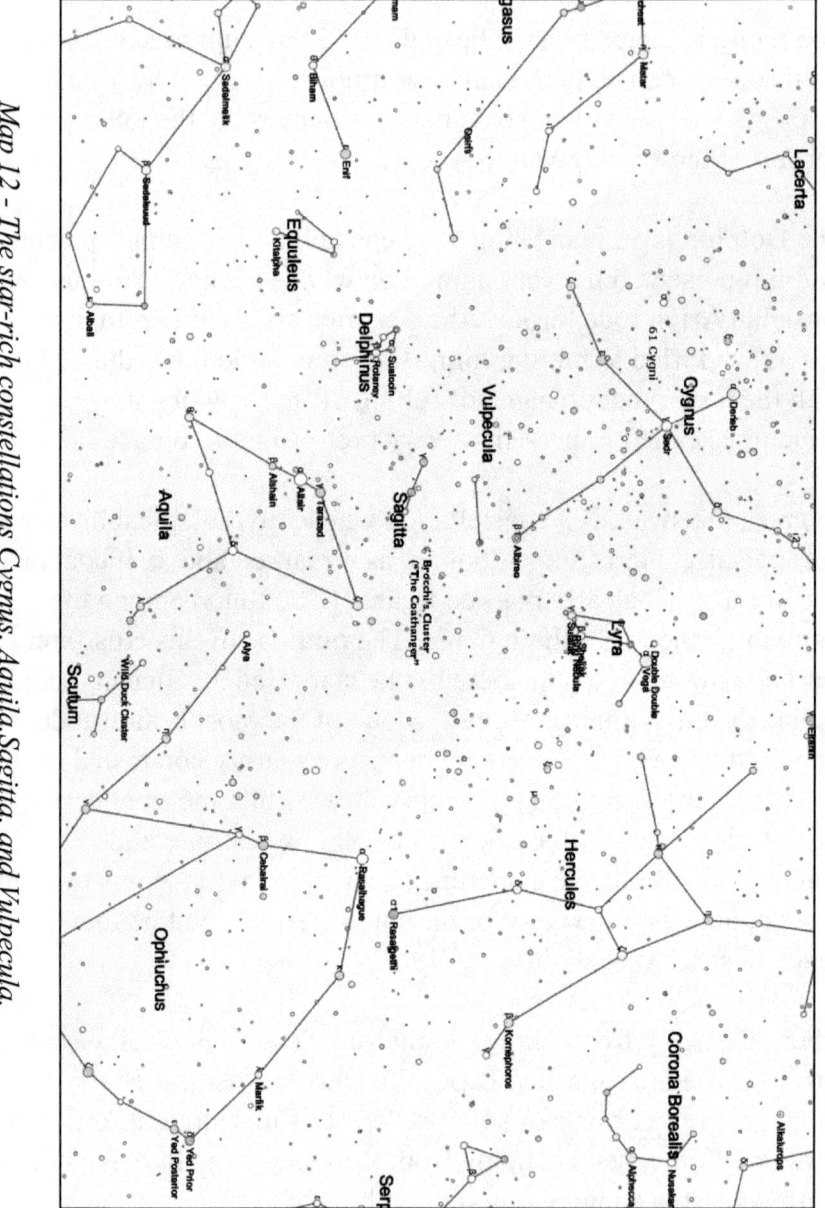

Map 12 - The star-rich constellations Cygnus, Aquila, Sagitta, and Vulpecula.

us, so far as we know, of all the stars in the northern hemisphere, and one of the nearest stars in all the heavens. Yet it is very faint, and the fact that so inconspicuous a star should be nearer than such brilliants as Vega and Arcturus shows how wide is the range of true brightness among the suns that light the universe. The actual distance of 61 Cygni is 11.4 light years. A modest glass will show this star about two binocular fields-of-view southeast of Deneb.

The star Omicron (o) Cygni is very interesting with binoculars. The naked eye sees a little star near it. The glass throws them wide apart into two colorful stars.

Now turn to the tail of Aquila, the Eagle. Just southwest of Lamda (λ) Aquilae, you will see the splendid open star cluster M11, also called the Wild Duck Cluster because of its appearance in a telescope. While it appears faint and unresolved in binoculars, a small telescope reveals its full splendor.

Sweep around Deneb and Sadr for the splendid star-fields that abound in this neighborhood; also around the upper part of the figure of the cross. We are here in one of the richest parts of the Milky- Way. Between the stars Deneb, Sadr, and Gienah, is the dark gap in the star
clouds of the galaxy, a sort of hole in the starry heavens caused by dust that lies between us and the distant stars. Although the region is not entirely empty of stars, its blackness is striking in contrast with the brilliancy of the Milky-Way in this neighborhood. The divergent streams of the great river of light in this region present a very remarkable appearance.

Draco, The Dragon

Finally, we come to the great dragon of the sky. In using the map of Draco and the neighboring constellations, the reader is supposed to face the north. The center of the upper edge of the map is directly over the observer's head. One of the stories told of this large constellation is that it represents a dragon that had the temerity to war against Minerva. The goddess *"seized it in her hand, and hurled it, twisted as it was, into the heavens round the axis of the world, before it had time to unwind its contortions."* Others say it is the dragon that guarded the golden apples in the Garden of the Hesperides, and that was slain by the redoubtable Hercules. At any rate, it is plainly a monster of the first magnitude.

The stars Etamin, Rastaban, Grumium, and Nu (v), represent its head, while its body runs trailing along, first sweeping in a long curve toward Cepheus, and then bending around and passing between the two bears. Try Nu with your binoculars, and if you succeed in seeing it double you may congratulate yourself on your keen sight. The distance between the stars is about 1 arcminute. Notice the contrasted colors of Etamin and Rastaban, the former being a rich orange and the latter white. As you sweep along the winding way that Draco follows, you will run across many striking fields of stars, although the heavens are not as rich here as in the splendid regions that we have just left.

You will also find that Cepheus, although not an attractive constellation to the naked eye, is worth some attention with binoculars. The head and upper part of the body of Cepheus are plunged in the stream of the Milky Way, while his feet are directed toward the pole of the heavens, upon which he is pictured as standing. One fine object for binoculars is the deep red star Mu (μ) Cephei, also called "The Garnet Star". It lies between Alderamin and Epsilon Cephei and is easily distinguishable by its deep red hue.

Cepheus, however, sinks into insignificance in comparison with its neighbor Cassiopeia, but that constellation belongs rather to the autumn sky, and we shall pass it by now.

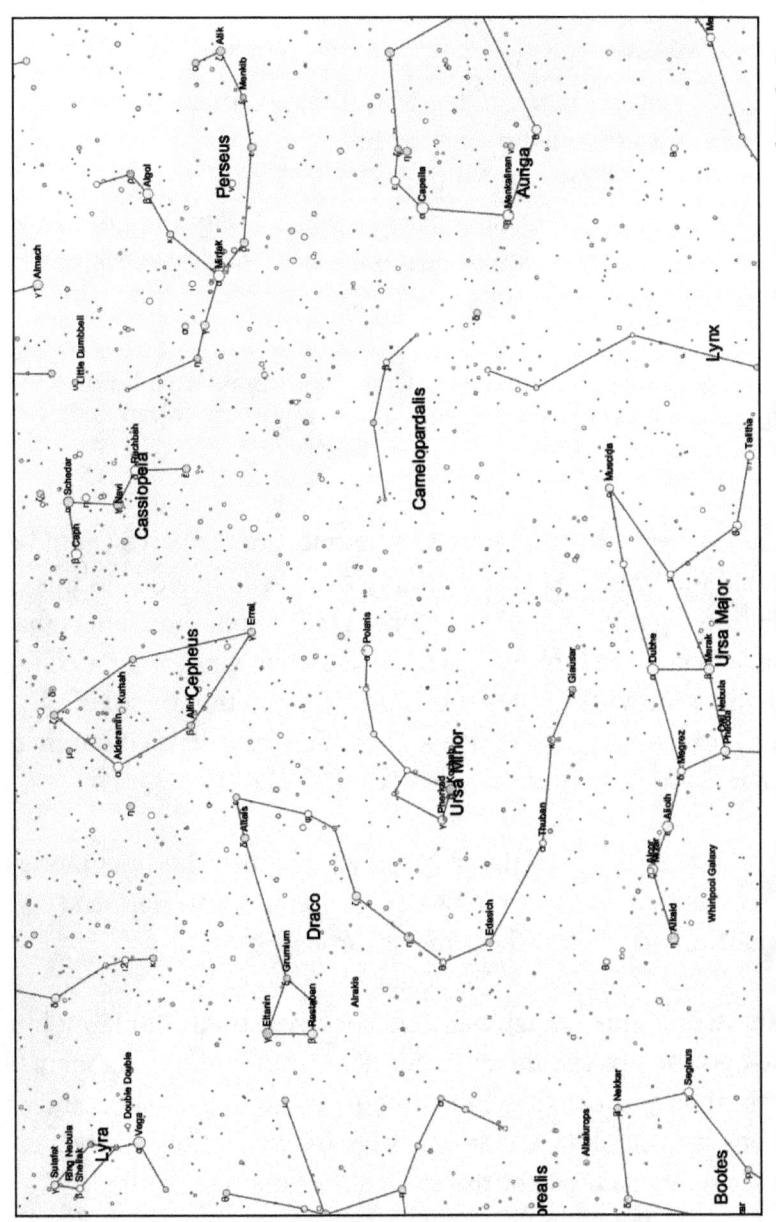

Map 13 - Looking directly overhead at Polaris, with Draco winding between the Great and Little Bears

81

The Stars of Autumn

A "Delightful Evening"

In a delightful, old, out-of-date book called the "*Plurality of Worlds*," the Astronomer and the Marquess, who have been making a wonderful pilgrimage through the heavens during their evening strolls in the park, come at last to the starry systems beyond the solar system and the Marquess experiences a lively impatience to know what the fixed stars will turn out to be, for the Astronomer has sharpened her appetite for marvels.

"Tell me," she says, eagerly, "are they, too, inhabited like the planets, or are they not peopled?In short, what can we make of them"

The Astronomer answers his charming questioner, as we should do today, that the fixed stars are so many suns. And he adds to this information a great deal of entertaining talk about the planets that may be supposed to circle around these distant suns, interspersing his conversation with explanations of the formation of solar systems and many quaint conceits, in which he is helped out by the ready wit of the Marquess.

Finally, the keen mind of the lady is overwhelmed by the grandeur of the scenes that the Astronomer opens to her view, her head swims, infinity oppresses her, and she cries for mercy.

"You show me," she exclaims, "a perspective so interminably long that the eye can not see the end of it. I see plainly the inhabitants of the earth; then you cause me to perceive those of the moon and of the other planets belonging to our solar system, quite clearly, yet not so distinctly as those of the earth. After them come the inhabitants of planets in the other solar systems. I confess, they seem to me hidden deep in the background, and, however hard I try, I can barely glimpse them at all. In truth, are they not almost annihilated by the very expression which you are obliged to use in speaking of them? You have to call them inhabitants of one of the planets contained in one of the infinity of solar systems. Surely we ourselves, to whom the same expression applies, are almost lost among so many millions of worlds. For my part, the earth begins to appear so frightfully little to me that henceforth I shall hardly consider any object worthy of eager pursuit. Assuredly, people who seek so earnestly their own aggrandizement, who lay schemes upon schemes, and give themselves so much trouble, know nothing of the infinity of the universe! I am sure my increase of knowledge will rebound to the credit of my idleness, and when people reproach me with indolence I shall reply : 'Ah ! if you but knew the history of the fixed stars!'"

It is certainly true that a contemplation of the unthinkable vastness of the universe, in the midst of which we dwell upon a speck illuminated by a spark, is calculated to make all terrestrial affairs appear contemptibly insignificant. We can not wonder that men for ages regarded the earth as the center, and the heavens with their lights as tributary to it, for to have thought otherwise, in those times, would have been to see things from the point of view of a superior intelligence. It has taken a vast amount of experience and knowledge to convince men of the insignificance of themselves and their belongings. So, in all ages they have applied a terrestrial measure to the universe, and imagined they could behold human affairs reflected in the heavens and human interests setting the gods together by the ears.

This is clearly shown in the story of the constellations.

The tremendous truth that on a starry night we look, in every direction, into an almost endless vista of suns beyond suns and systems upon systems, was too overwhelming for comprehension by the inventors of the constellations. So they amused themselves, like imaginative children, as they were, by tracing the outlines of men and beasts formed by those pretty lights, the stars. They

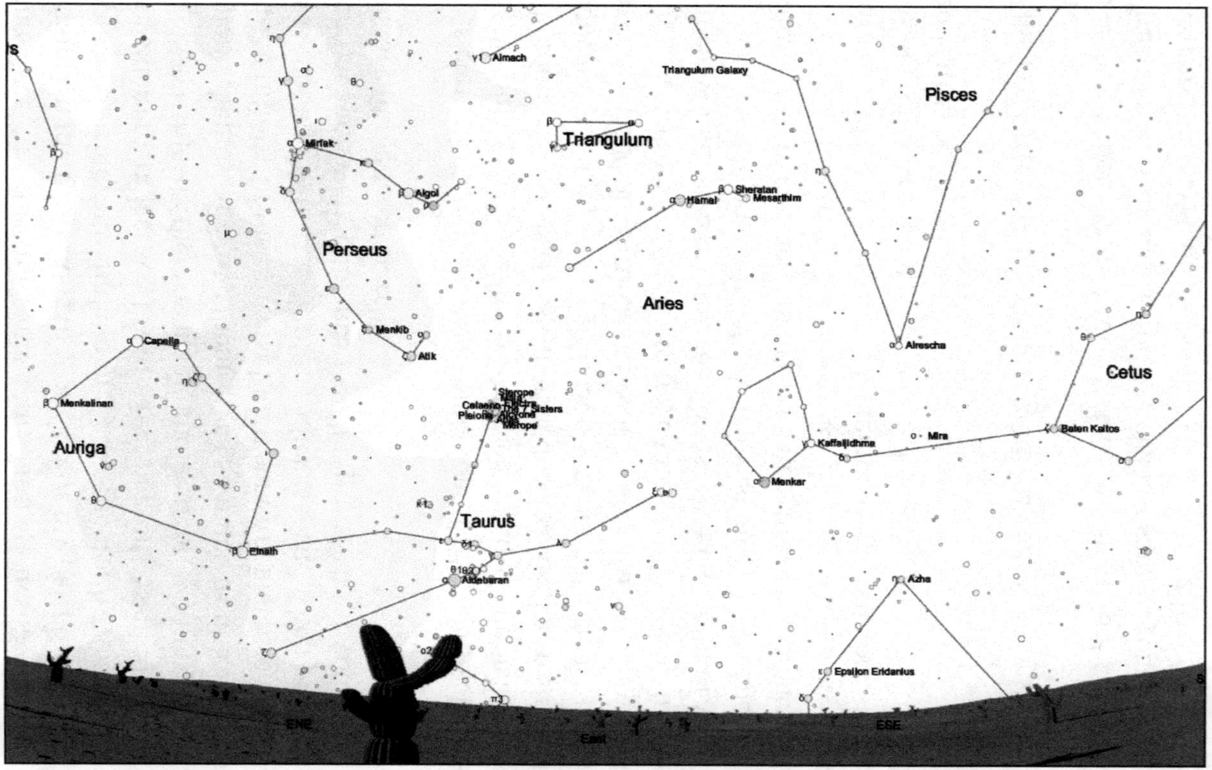

Map 14a - The Autumn Constellations (looking EAST at 40 degrees N latitude, at 9 p.m. on Oct 15)

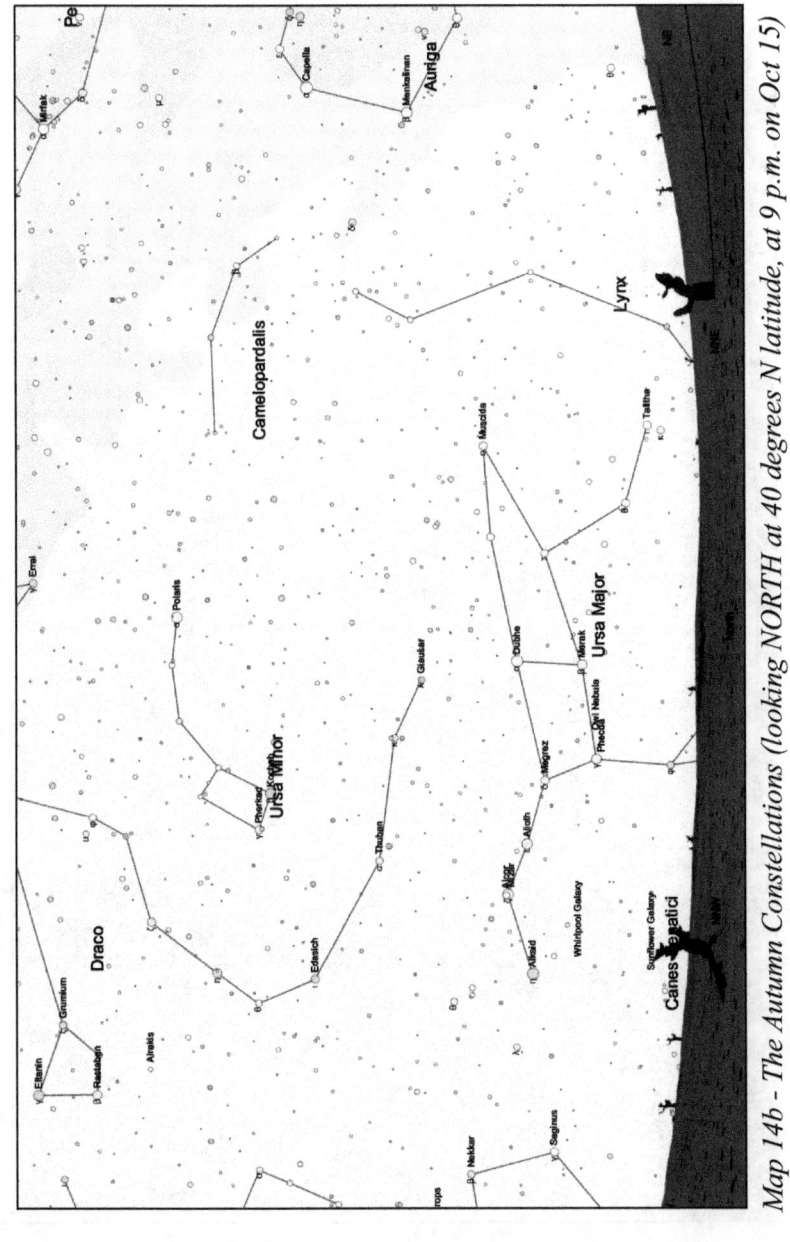

Map 14b - The Autumn Constellations (looking NORTH at 40 degrees N latitude, at 9 p.m. on Oct 15)

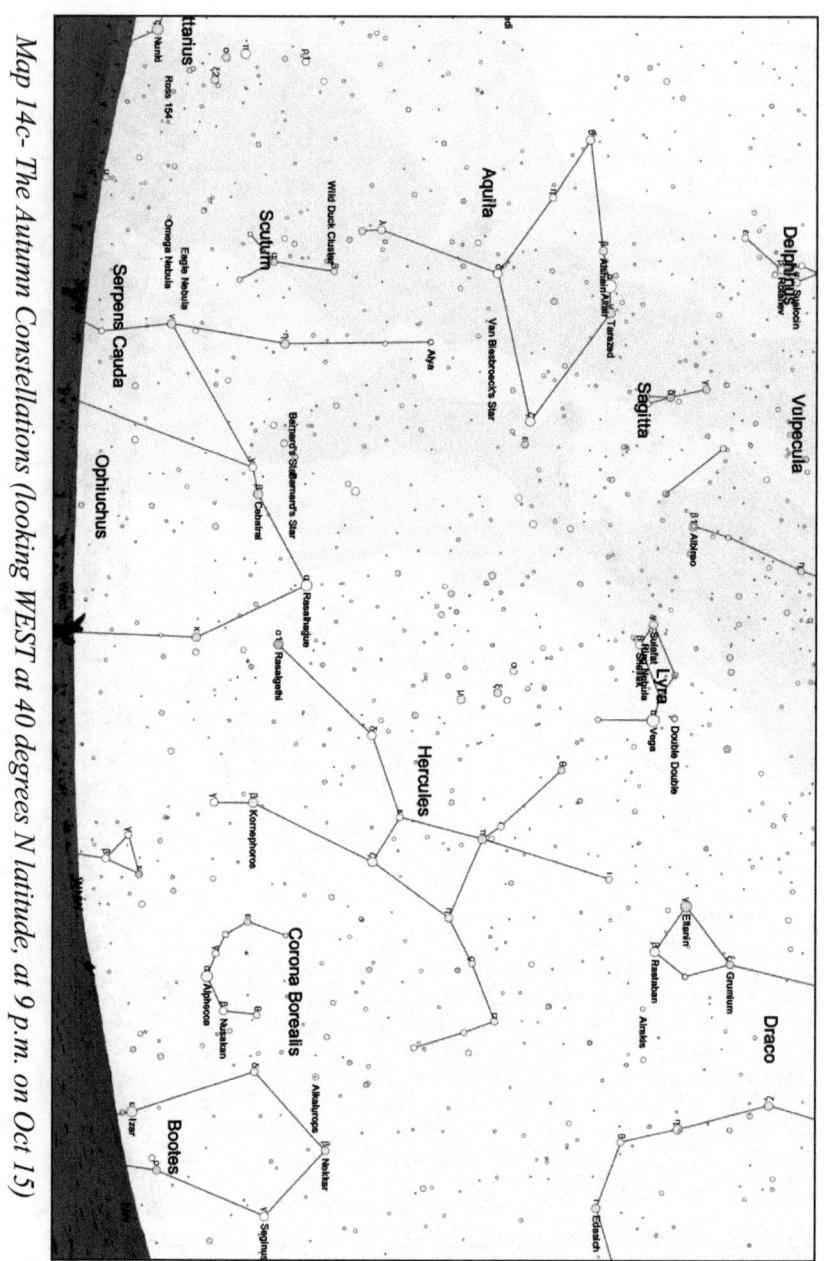

Map 14c- The Autumn Constellations (looking WEST at 40 degrees N latitude, at 9 p.m. on Oct 15)

86

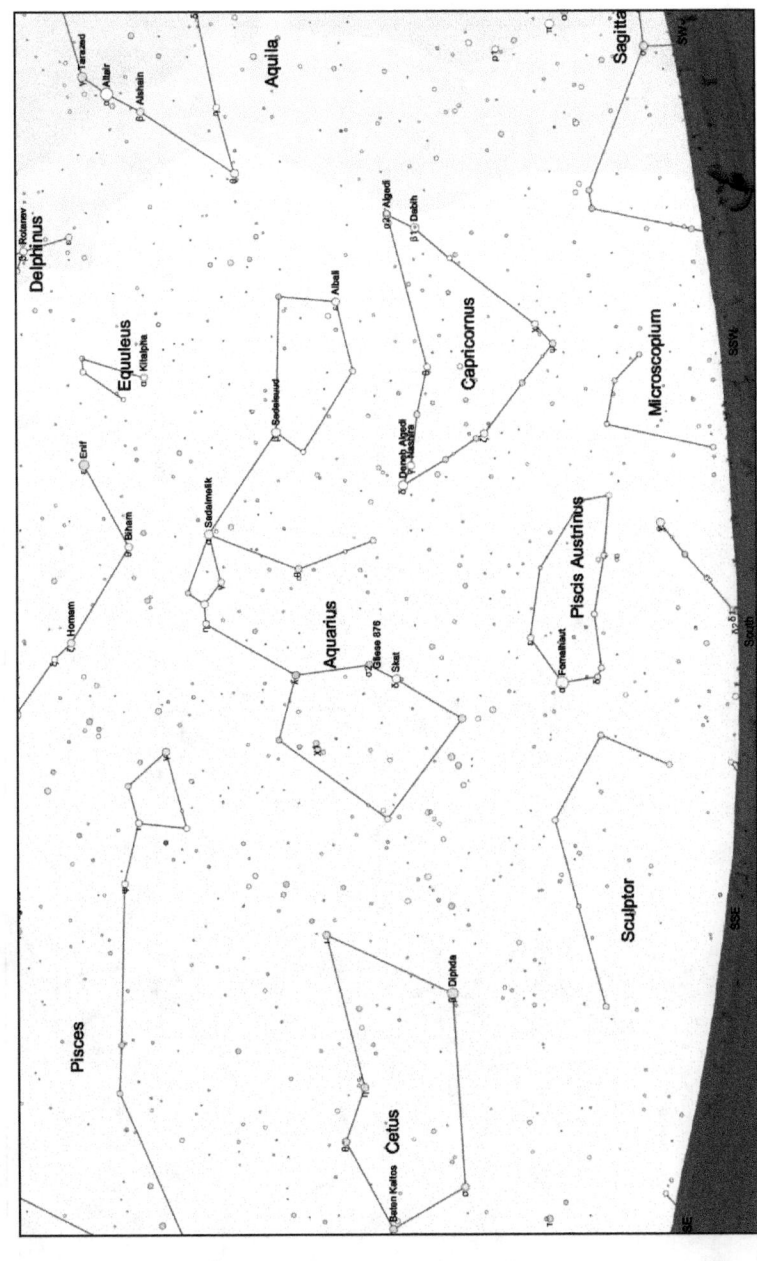

Map 14d- The Autumn Constellations (looking SOUTH at 40 degrees N latitude, at 9 p.m. on Oct 15)

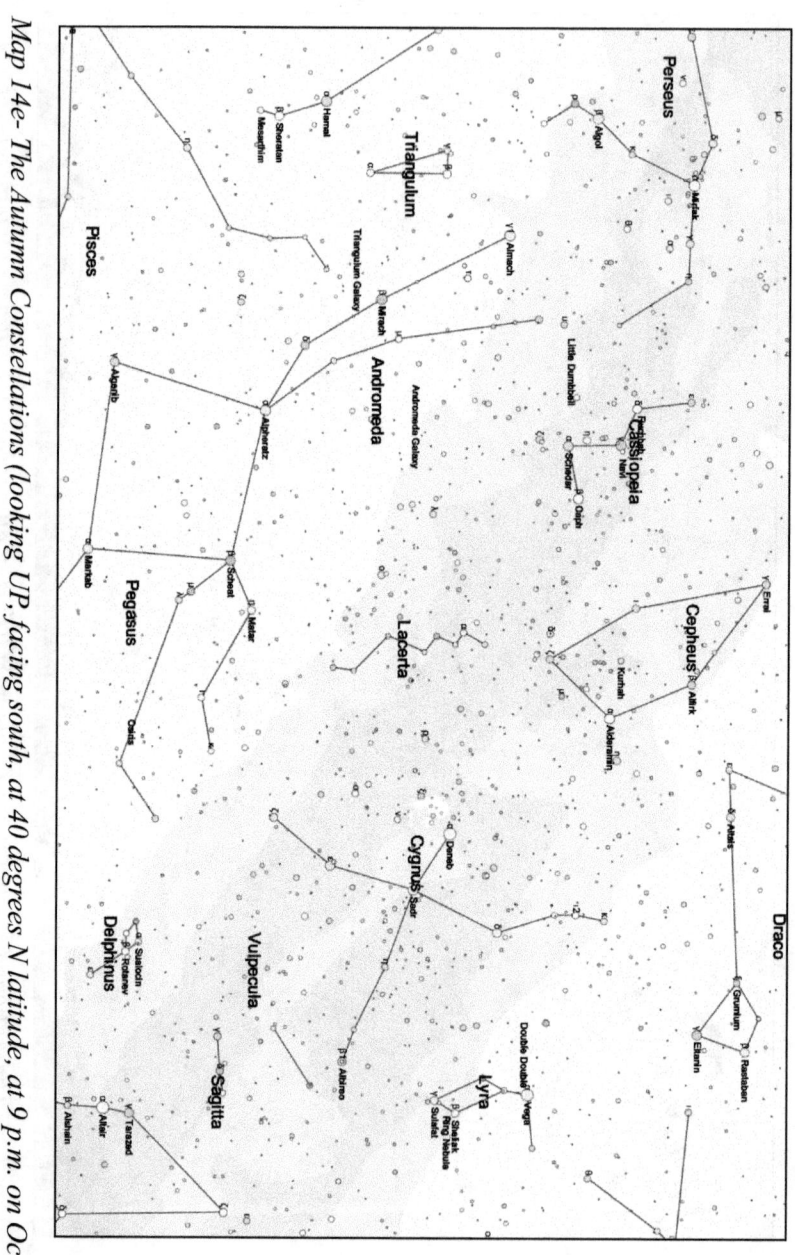

Map 14e- The Autumn Constellations (looking UP, facing south, at 40 degrees N latitude, at 9 p.m. on Oct 15)

turned the starry heavens into a scroll filled with pictured stories of mythology.

Four of the constellations with which we are going to deal in this chapter are particularly interesting on this account. They preserve in the stars, more lasting than parchment or stone, one of the oldest and most pleasing of all the romantic stories that have amused and inspired the minds of men the story of Perseus and Andromeda-- a

The attendants of alpha Persei (Mirphak)

better story than any that modern novelists have invented. The four constellations to which I refer bear the names of Andromeda, Perseus, Cassiopeia, and Cepheus, and are sometimes called, collectively, the Royal Family. In the autumn they occupy a conspicuous position in the sky, forming a group that remains unrivaled until the rising of Orion with his imperial *cortége*. The reader will find them in Map 14, occupying the northeastern quarter of the heavens.

This map represents the visible heavens at about midnight on September 1st, 10 o'clock p.m. on October 1st, and 8 o'clock p.m. on November 1st. At this time the constellations that were near the meridian in summer will be found sinking in the west, Hercules being low in the northwest, with the brilliant Lyra and the head of Draco suspended above it; Aquila, "the eagle of the winds,"soars high in the southwest ; while the Cross of Cygnus is just west of the zenith; and Sagittarius, with its wealth of star-dust, is disappearing under the horizon in the southwest.

Capricorn, and Piscis Australis

Far down in the south the observer catches the gleam of a bright lone star of the first magnitude, though not one of the largest of that class. It is Fomalhaut, in the mouth of the Southern Fish, Piscis Australis. A slight reddish tint will be perceived in the light of this beautiful star, whose brilliance is enhanced by the fact that it shines without a rival in that region of the sky. Fomalhaut is one of the important "nautical stars," and its position was long ago carefully computed for the benefit of mariners. The constellation of Piscis Australis, which will be found in our second map, does not possess much to interest us except its splendid leading star. In consulting Map 15, the observer is supposed to be facing south, or slightly west of south, and he must remember that the upper part of the map reaches nearly to the zenith, while at the bottom it extends down to the horizon.

To the right, or west, of Fomalhaut, and higher up, is the constellation of Capricornus, very interesting on many accounts, though by no means a striking constellation to the unassisted eye. The stars Alpha (α), called Giedi, and Beta (β), called Dabih, will be readily recognized, and a keen eye will perceive that Alpha really consists of two stars. They are about six minutes of arc apart, and are of the third and the fourth magnitude respectively. These stars, which to the naked eye appear almost blended into one, really have no physical connection with each other, and are slowly drifting apart. The ancient astronomers make no mention of Giedi being composed of two stars, and the reason is plain: in the time of Hipparchus, their distance apart was not more than two thirds as great as it is at present, so that the naked eye could not have detected the fact that there were two of them; and it was not until the seventeenth century that they got far enough asunder to begin

90

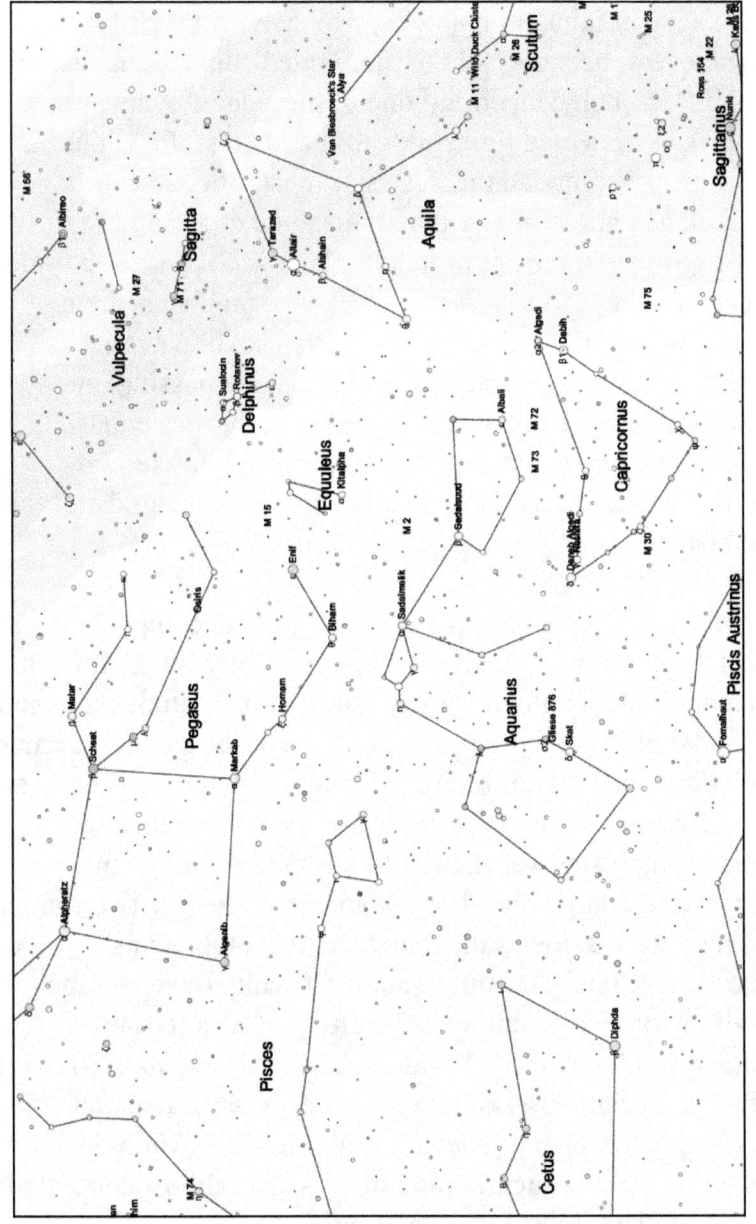

Map 15 - The grand constellations Capricorn, Pegasus, Aquarius, and Piscis Austrinus

to be separated by eyes of unusual power. With binoculars, they are thrown well apart, and present a very pretty sight. Considering the manner in which these stars are separating, the fact that both of them have several faint companions, which our powerful telescopes reveal, becomes all the more interesting. A suggestion of Sir John Herschel, concerning one of these faint companions, that it shines by reflected light, adds to the interest, for if the suggestion is well founded the little star must, of course, be actually a planet, and granting that, then some of the other faint points of light seen there are probably planets too. It must be said that Herschel's suggestion was incorrect. The faint stars are not planets, but shine with their own light. Even so, however, these two systems, which apparently have met and are passing one another, at a distance small as compared with the space that separates them from us, possess a peculiar interest, like two celestial fleets that have spoken to one another in the midst of the ocean of space.

The star Beta, or Dabih, is also a double star. The companion is of a beautiful blue color, generally described as "sky-blue." It is of the seventh magnitude, while the larger star is of magnitude three and a half. The latter is golden-yellow. The blue of the small star can be seen with binoculars, but it requires careful looking and a clear and steady atmosphere. I found the color was even more distinct, although the small star was then more or less enveloped in the yellow rays of the large one. The distance between the two stars in Dabih is nearly the same as that between the components of Epsilon (ε) Lyrae, and the comparative difficulty of separating them is an instructive example of the effect of a large star in concealing a small one close beside it. The two stars in ε Lyrae are of nearly equal brightness, and are very easily separated and distinguished but in Dabih, one star is about twenty times as bright as the other, and consequently the fainter star is almost concealed in the glare of its more brilliant neighbor.

Now sweep from the star Zeta (ζ) eastward a distance somewhat greater than that separating Alpha and Beta, and you will find a fifth-magnitude star beside a little nebulous spot. This is the cluster known as M30, one of those sun-swarms that over- whelm the mind of the contemplative observer with astonishment, and especially remarkable in this case for the apparent vacancy of the heavens immediately surrounding the cluster, as if all the stars in that neighborhood had been drawn into the great assemblage, leaving a void around it. Of course, with the instrument that our observer is supposed to be using, merely the existence of this lovely globular cluster, much like the Great Cluster of Hercules, can be detected ; but, if he sees that it is there, he may be led to provide himself with a telescope capable of revealing its glories.

Admiral Smyth remarks that, "although Capricorn is not a striking object, it has been the very pet of all constellations with astrologers. The mythological account of the constellation is that it represents the goat into which Pan was turned in order to escape from the giant Typhon, who once upon a time scared all the gods out of their wits, and caused them to change themselves into animals, even Jupiter assuming the form of a ram. According to some authorities, Piscis Australis represents the fish into which Venus changed herself on that interesting occasion.

Aquarius

Directly above Piscis Australis, and to the east or left of Capricorn, the map shows the constellation of Aquarius, or the Water-Bearer. Some say this commemorates Ganymede, the cup-bearer of the gods. It is represented in old star-maps by the figure of a young man pouring water from an urn. The star Sadalmelik marks his right shoulder, and Sadalsuud his left, and Sadachbia, Zeta (ζ), Eta (η), and Pi (π) indicate his right hand and the urn. From this group a current of small stars will be recognized, sweeping downward with a curve toward the east, and ending at Fomalhaut; this represents the water poured from the urn, which the Southern Fish

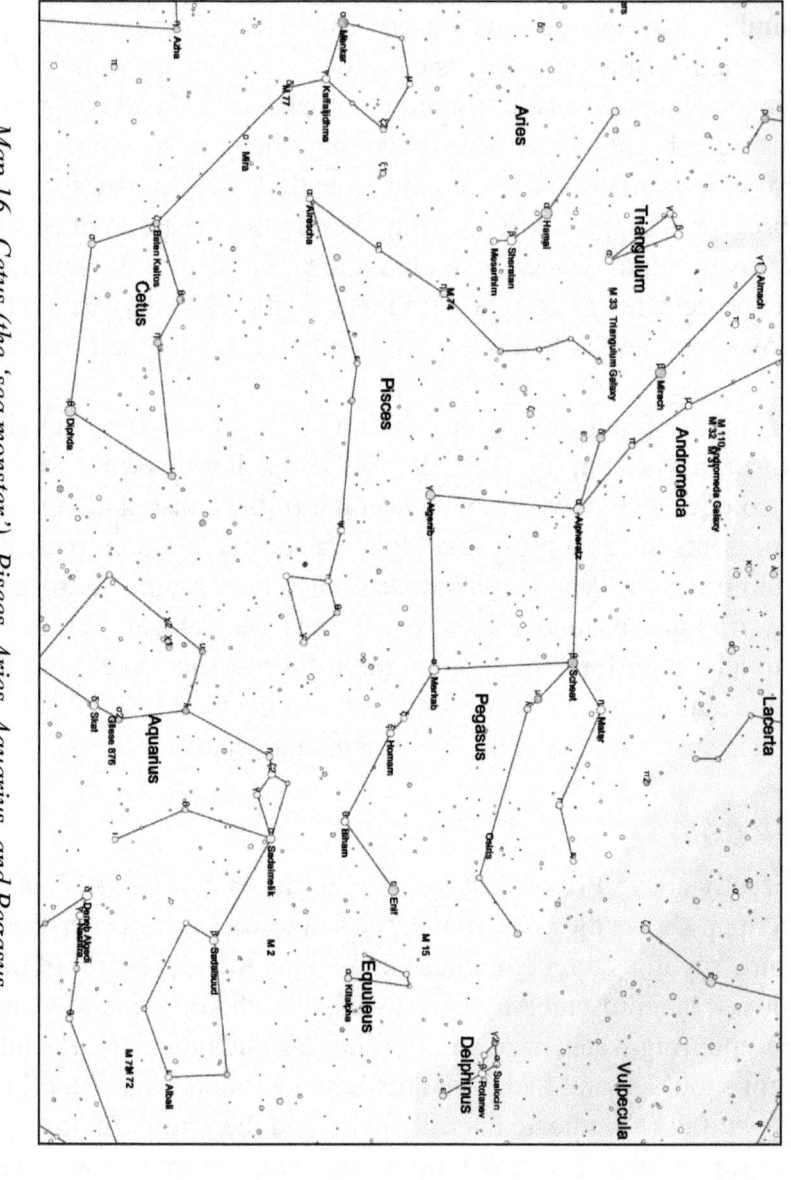

Map 16 - Cetus (the 'sea monster'), Pisces, Aries, Aquarius, and Pegasus.

appears to be drinking. In fact, according to the pictures in the old maps, the fish succeeds in swallowing the stream completely, and it vanishes from the sky in the act of entering his distended mouth! In Greek, Latin, and Arabic this constellation bears names all of which signify "a man pouring water." The ancient Egyptians imagined that the setting of Aquarius caused the rising of the Nile, as he sank his huge urn in the river to fill it. Alpha was called by the Arabs Sadalmelik, which is interpreted to mean the "king's lucky star," but whether it proved itself a lucky star in war or in love, and what particular king enjoyed its benign influence and recorded his gratitude in its name, we are not told. Thus, at every step, we find how shreds of history and bits of superstition are entangled among the stars. Surely, humanity has been reflected in the heavens as lastingly as it has impressed itself upon the earth.

Starting from the group of stars just described as forming the Water-Bearer's urn, follow with a glass the winding stream of small stars that represent the water. Several very pretty and striking assemblages of stars will be encountered in its course. The star Tau (τ) is double and presents a beautiful contrast of color, one star being white and the other reddish-orange.

Point a good glass upon the star marked Nu (ν), and you will see, somewhat less than a degree and a half to the west of it, what appears to be a faint star of between the seventh and eighth magnitudes. You will have to look sharp to see it. Pristine dark sky is essential. The faint speck is a planetary nebula, unrivaled for interest by many of the larger and more conspicuous objects of that kind. It resembles the planet Saturn; in other words, that it consists apparently of a ball surrounded by a ring. But the spectroscope proves that it is a gaseous mass, and its size is sufficient to fill the orbit of Neptune! The shape of the Saturn Nebula proclaims unmistakably the operation of great metamorphic forces there. This planetary nebula, like others of its kind such as the Ring Nebula in Lyra, is the consequence of a dying star which is throwing off its outer layers into space. Of course, with binoculars, you can see nothing of the strange form of this object, the detection of which requires the aid of powerful telescopes, but it is much to know where that unfinished destruction lies, and to see it, even though diminished by distance to a mere speck of light.

Further westward by a degree, and slightly southward, you will see M73, which was once believed to be a cluster of young stars, but which is now know to be only a chance alignment of stars in space. And a degree further to the west, you will glimpse the fuzzy speck of another globular cluster M72.

Turn your glass upon the star shown in the map just above Mu (μ) and Epsilon (ε). You will find an attractive arrangement of small stars in its neighborhood. One is double to the naked eye, and the row of stars below it is well worth looking at. The star Delta (δ), called Skat, indicates the place where, in 1756, Tobias Mayer narrowly escaped making a discovery that would have anticipated that which a quarter of a century later made the name of Sir William Herschel world-renowned. The planet Uranus passed near Delta in 1756, and Tobias Mayer saw it, but it moved so slowly that he took it for a fixed star, never suspecting that his eyes had rested upon a member of the solar system whose existence was, up to that time, unknown to the inhabitants of this world.

Pegasus, The Winged Horse

Above Aquarius you will find the constellation Pegasus. It is conspicuously marked by four stars of about the second magnitude, which shine at the corners of a large square, called the Great Square. This figure is some fifteen degrees square, and at once attracts the eye, there being few stars visible within the quadrilateral, and no large ones in the immediate neighborhood to distract attention from it. One of the four stars, however, as will be seen by consulting Map 15, does not belong to Pegasus, but to the constellation Andromeda. Mythologically, this constellation represents the celebrated winged horse of antiquity.

The star Alpha (α) is called Markab; Beta (β) is Scheat, and Gamma (γ) is Algenib; the fourth star in the square, belonging to Andromeda, is called Alpheratz. Although Pegasus presents a striking appearance to the unassisted eye, on account of its great square, it contains little to attract the observer with small binoculars. An exception is the fine globular cluster M15, which resembles the Great Cluster in Hercules, but presents a tighter star-like structure than M13. Make no mistake, it is a majestic cluster of hundreds of thousands of ancient stars. You will easily see M15 in dark sky off the nose of Pegasus by following the line extending from the stars Biham to Enif.

It will also prove interesting, however, to sweep with the glass carefully over the space within the square, which is comparatively barren to the naked eye, but in which many small stars will be revealed, of whose existence the naked-eye observer would be unaware. The star marked Pi Peg (π), in the horse's foot, is an interesting double which can be separated by a good eye without artificial aid, and which, with binoculars, presents a fine appearance.

Cetus, The Sea Monster

And now we come to Map 16, representing the constellations Cetus, Pisces, Aries, and Triangulum. In consulting it, you are supposed to face the southeast. Cetus is a very large constellation, and from the peculiar conformation of its principal stars it can be readily recognized. The head is to the east, the star Alpha (α), called Menkar, lies in the nose of this imaginary inhabitant of the sky-depths. The constellation is supposed to represent the monster that, according to fable, was sent by Neptune to devour the fair Andromeda, but whose bloodthirsty plan was gallantly frustrated by Perseus, as we shall learn from starry mythology further on.

Although bearing the name Cetus, the Whale, the pictures of the constellation in the old maps do not present us with the form of a whale, but that of a most extraordinary scaly creature with enormous jaws filled with large teeth, a forked tongue, forepaws armed with gigantic claws, and a long, crooked, and dangerous-looking tail. Indeed, Aratus does not call it a "whale," but a "sea-monster," whose terrible prowess is celebrated in the book of Job.

By far the most interesting object in Cetus is the star Mira. This is a famous variable star that sometimes shines more than a thousand-fold more brilliantly than at others! It changes from the second magnitude to the ninth or tenth, its period from maximum to maximum being about eleven months. During about five months of that time it is completely invisible to the naked eye; then it begins to appear again, slowly increasing in brightness for some three months, until it shines as a star of the second or third magnitude, being then as bright as, if not brighter than, the most brilliant stars in the constellation. It retains this brilliance for about two weeks, and then begins to fade again, and, within a few months, once more disappears. There are various irregularities in its changes, which render its exact period somewhat uncertain, and it does not always attain the same degree of brightness at its maximum. For instance, in 1779, Mira was almost equal in brilliance to a first-magnitude star, but frequently at its greatest brightness it is hardly equal to an ordinary star of the second magnitude. By the aid of our little map you will readily be able to find it. You will perceive that it has a slightly reddish tint. Watch it from one of its maxima, and you will see it gradually fade from sight until, at last, only the blackness of the empty sky appears where, a few months before, a conspicuous star was visible. Keep watch of that spot, and in due course you will perceive Mira shining there again a mere speck, but slowly brightening and in three months more the wonderful star will blaze again with renewed splendor.

Mira's variations arise from a curious and unstable pulsating of its atmosphere, as the pull of gravity and the unsteady burning of fuel in its core struggle for influence. Each dominates for a few months before the other reasserts itself. In the course of pulsation, much of the star's outer layers are lost forever into space, perhaps to seed the formation of new stars and worlds. Mira presents to us an example of what our sun will be in the course of time, as our star ages and slowly evolves towards its inevitable demise.

There are several other variable stars in Cetus, but none possessing much interest for us. The observer should look at the group of stars in the head, where he will find some interesting combinations, and also at Chi (χ), which is the little star shown in the map near Zeta (ζ). This is a double that will serve as a very good test of eye and instrument, the smaller companion-star being of only seven and a half magnitude.

Pisces, Aries, and Triangulum

Directly above Cetus is the long, straggling constellation of Pisces, the Fishes. The Northern Fish is represented by the group of stars near Andromeda. A long band or ribbon, supposed to bind the fish together, trends first southeast and then west until it joins a group of stars under Pegasus, which represents the Western Fish, not to be confounded with the Southern Fish described near the beginning of this chapter, which is a separate constellation. Fable has, however, somewhat confounded these fishes; for while, as I have remarked above, the Southern Fish is said to represent Venus after she had turned herself into a fish to escape from the giant Typhon, the two fishes of the constellation we are now dealing with are also fabled to represent Venus and her interesting son Cupid under the same disguise assumed on precisely the same occasion. If Typhon, however, was so great a brute that even Cupid's arrows were of no avail against him, we should, perhaps, excuse mythology for duplicating the record of so wondrous an event.

You will find it very interesting to take your glass and, beginning with the attractive little group in the Northern Fish, follow the windings, of the ribbon, with its wealth of tiny stars, to the Western Fish. When you have arrived at that point, sweep well over the sky in that neighborhood, and particularly around and under the stars Iota (ι), Theta (θ), Lambda (λ), and Gamma (γ). If you are using a powerful glass, you will be surprised and delighted by what you see. Below the star Omega (ω), and to the left of Lambda, is the place which the sun occupies at the time of the spring (or vernal) equinox in other words, one of the two crossing-places of the equator of the heavens, and the ecliptic, or the sun's path. The prime meridian of the heavens passes through this point. You can trace out this great circle, from which astronomical longitudes are reckoned, by drawing an imaginary line from the point of the vernal equinox just indicated through Alpha (α) in Andromeda and Beta (β) in Cassiopeia to the Polaris, the North Star.

To the left of Pisces, and above the head of Cetus, is the constellation Aries, or the Ram. Two pretty bright stars, four degrees apart, one of which has a fainter star near it, mark it out plainly to the eye. These stars are in the head of the Ram. The brightest one, Alpha (α), is called Hamal ; Beta (β) is named Sheratan; and its fainter neighbor Gamma (γ) is Mesartim. According to fable, this constellation represents the ram that wore the golden fleece, which was the object of the celebrated expedition of the Argonauts. There is not much in the constellation to interest us, except its historical importance, as it was more than two thousand years ago the leading constellation of the zodiac, and still stands first in the list of the zodiacal signs. Owing to the precession of the equinoxes, however, the vernal equinox, which was formerly in this constellation, has now advanced into the constellation Pisces, as we saw above. Mesartim is interesting as the first telescopic double star ever discovered. It was detected by Robert Hooke while watching the passage of a comet near the star in 1664. Singularly enough, the brightest star in the constellation,

Hamal, originally did not belong to the constellation. Tycho Brahe finally placed it in the head of Aries.

Next, to the little constellation of Triangulum, just above Aries. Insignificant as it appears, this little group is a very ancient constellation. It received its name, Deltoton, from the Greek letter Delta (capitalized as Δ).

A keen observer in very dark sky will observe the delightful face-on spiral galaxy M33, some 3 degrees west of the star Mothallah. Spread out over an area larger than the full moon, this galaxy is quite faint in binoculars, though some can see it without optical aid. M33 is associated gravitationally with our own Milky Way and with the magnificent Andromeda Galaxy, to which we shall turn shortly. The light you see from this galaxy left its constituent stars some 3 million years ago.

Perseus, Andromeda, and Cassiopeia

You will now be introduced to the "Royal Family." Although the story of Perseus and Andromeda is, of course, well known to many readers, yet, on account of the great beauty and brilliancy of the group of constellations that perpetuate the memory of it among the stars, it is worth recalling here.

As Perseus was returning through the air on the back of Pegasus from his conquest of the Gorgon Medusa, he saw the beautiful Andromeda chained to a rock on the sea-coast, waiting to be devoured by a sea-monster. The poor girl's only offense was that her mother, Cassiopeia, had boasted for her that she was fairer than the the fifty sea-nymphs, the daughters of the gods Nereus and Doris. For this Neptune had decreed that all the land of the Ethiopians should be drowned and destroyed unless Andromeda was delivered up as a sacrifice to the dreadful sea-monster. When Perseus, dropping down to learn why this maiden was chained to the rocks, heard from Andromeda's lips the story of her woes, he laughed with joy. Here was an adventure just to his liking, and besides, unlike his previous adventures, it involved the fate of a beautiful woman with whom he was already in love. Could he save her ? Well, wouldn't he! The sea-monster might frighten an entire kingdom, but it could not shake the nerves of the hero from Greece. He whispered words of encouragement to Andromeda, who could scarce believe the good news that a champion had come to defend her after all her friends and royal relations had deserted her. Neither could she feel much confidence in her young champion's powers when suddenly her horrified gaze met the awful leviathan of the deep advancing to his feast! But Perseus, with a warning to Andromeda not to look at what he was about to do, sprang with his winged sandals up into the air. And then...

"On came the great sea-monster, coasting along like a huge black galley, lazily breasting the ripple, and stopping at times by creek or headland to watch for the laughter of girls at their bleaching, or cattle pawing on the sand-hills, or boys bathing on the beach. His great sides were fringed with clustering shells and sea-weeds, and the water gurgled in and out of his wide jaws as he rolled along, dripping and glistening in the beams of the morning sun. At last he saw Andromeda, and shot forward to take his prey, while the waves foamed white behind him, and before him the fish fled leap- ing.

"Then down from the height of the air fell Perseus like a shooting-star down to the crest of the waves, while Andromeda hid her face as he shouted. And then there was silence for a while.

"At last she looked up trembling, and saw Perseus springing toward her ; and, instead of the monster, a long, black rock, with the sea rippling quietly round it."

Perseus had turned the monster into stone by holding the blood-freezing head of Medusa before his eyes ; and it was fear lest

Andromeda herself might see the Gorgon's head, and suffer the fate of all who looked upon it, that had led him to forbid her watching him when he attacked her enemy. Afterward he married her, and Cassiopeia, Andromeda's mother, and Cepheus, her father, gave their daughter's rescuer a royal welcome, and all rose up and blessed him for ridding the land of the monster. And now, if we choose, we can, any fair night, see the principal characters of this old romance shining in the starry sky.

The makers of old star-maps seem to have vied in the effort to represent with effect the figures of Andromeda, Perseus, and Cassiopeia among the stars, and it must be admitted that some of them succeeded in giving no small degree of life and spirit to their sketches.

The starry riches of these constellations are well matched with their high mythological repute. Lying in and near the Milky-Way, they are particularly interesting to the observer with binoculars. Besides, they include several of the most celebrated wonders of the firmament.

In consulting Map 17, the observer is supposed to face the east and northeast. We will begin our survey with Andromeda. The three chief stars of this constellation are of the second magnitude, and lie in a long, bending row, beginning with Alpha (α), or Alpheratz, in the head, which, as we have seen, marks one corner of the great Square of Pegasus. Beta (β), or Mirach, with the smaller stars Mu (μ) and Nu (ν), form the waist. The third of the chief stars is Gamma (γ), or Almaak, situated in the left foot. The little group of stars designated Lambda (λ), Kappa (κ), and Iota (ι), mark the extended right hand chained to the rock, and Zeta (ζ) and some smaller stars southwest of it show the left arm and hand, also stretched forth and shackled.

In searching for picturesque objects in Andromeda, begin with Alpheratz and the groups forming the hands. Below the waist will be seen a rather remarkable arrangement of small stars in the mouth of the Northern Fish. Now follow up the line of the girdle to the star Nu (ν). If your glass has a wide field, your eye will immediately catch the glimmer of the Great Nebula of Andromeda in the same field with the star. This is the oldest or earliest discovered of the nebulae, and, with the exception of that in Orion, is the grandest visible in this hemisphere. Of course, not much can be expected in viewing such an object with binoculars; and yet a good glass, in clear weather and the absence of the moon, makes a very attractive spectacle of it. Of course, the nebula is in fact a galaxy much like own own Milky Way, though somewhat larger. At a distance of 2 million light years, the Andromeda Galaxy is the closest major galaxy to our own, and hosts some 300 billion stars. The expanse of our neighboring galaxy is such that the light from the near side left 150,000 years earlier than the light from the far side.

By turning the eyes aside, the galaxy can be seen, extended as a faint, wispy light, much elongated on either side of the brighter nucleus. The image above shows, approximately, the appearance of the nebula, together with some of the small stars in its neighborhood, as seen with binoculars. With large telescopes it appears both larger and broader, expanding to a truly enormous extent, and in photographs of it we behold gigantic rifts of dust running lengthwise, while the whole field of sky in which it is contained appears sprinkled over with minute stars which lie in the foreground in our galaxy. The Andromeda galaxy has two satellite galaxies, M32 and M110, but they require a large pair of binoculars or a telescope to be perceived.

It will be found very interesting to sweep with the glass slowly from side to side over Andromeda, gradually approaching toward Cassiopeia or Perseus. The increase in the richness of the stratum of faint stars that apparently forms the background of the sky will be clearly discernible as you approach the Milky Way, which passes directly through Cassiopeia and Perseus. The Milky-Way itself, in that splendidly rich region about Sagittarius (described in the "Stars of Summer"), is not nearly so spectacular a sight with binoculars as it is above Cygnus and in the region with which we are now dealing.

The star Nu, which serves as a pointer to the Great Nebula, is itself worth some attention with a strong glass on account of a pair of small stars near it.

The star Gamma (γ) is interesting, not only as one of the most beautiful triples in the heavens (binoculars are far too feeble an instrument to reveal its companions), but because it serves to indicate the radiant point of the Biela meteors. There was once a comet well known to astronomers by the name of its discoverer, Biela. It repeated its visits to the neighborhood of the sun once in every six or seven years. In 1846 this comet astonished all observers by splitting into two comets, which continued to run side by side, like two equal racers, in their course around the sun. Each developed a tail of its own. In 1852, when the twin comets were due again, the astronomical world was on watch, and the comets did not disappoint expectation, for back they came out of the depths of space, still racing, but much farther apart than they had been before, alternating in brightness as if the long struggle had nearly exhausted them, and finally, like spent runners, growing faint and disappearing. They have never been seen since.

In 1872, when the comets should have been visible, if they still existed, a very startling thing happened. Out of the northern heavens, along the track of the missing comets, where the earth crossed it, on the night of the 27th of November came glistening and dashing the fiery spray of a storm of meteors. It was the dust and fragments of the lost comet of Biela, which, after being split in two in 1852, had evidently continued the process of disintegration. It seems to have made a truly ghostly exit, for right after the meteor shower of 1872 a mysterious cometary body was seen, which was supposed at the time to be the missing comet itself, and

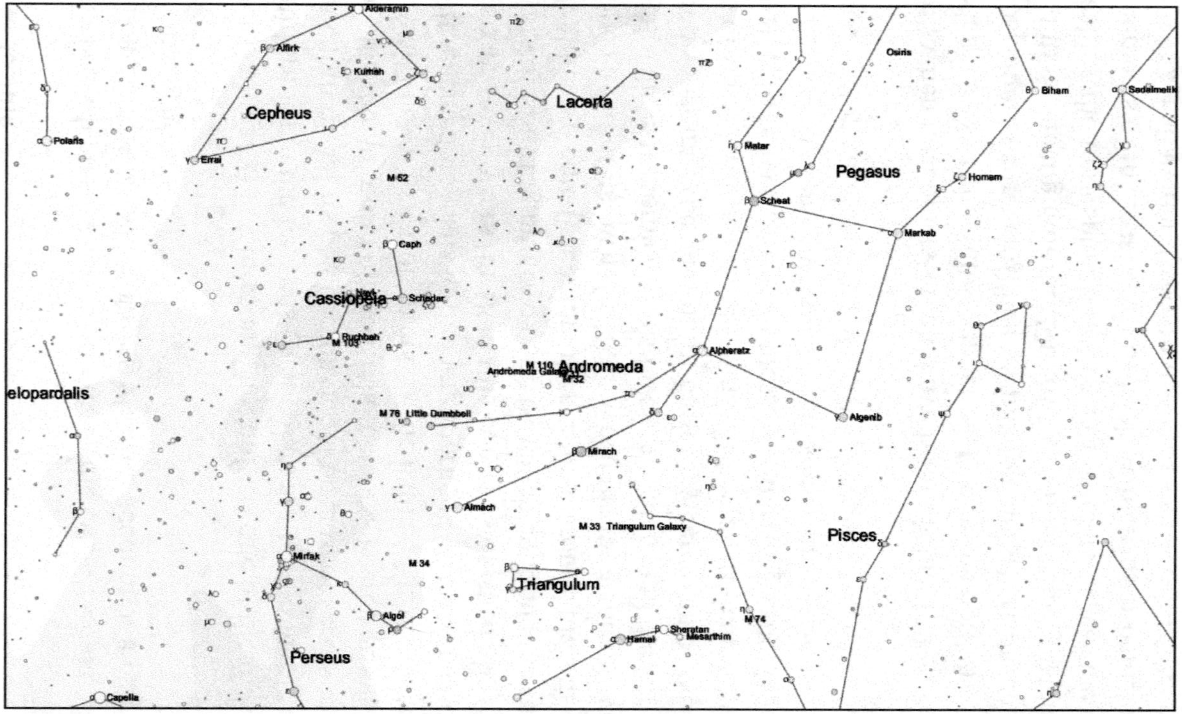

Map 17 - The "Royal Family" of constellations: Cepheus, Cassiopeia, Andromeda, and Perseus.

which, it is not altogether improbable, may have been a fragment of it. Three days after the meteors burst over Europe, German astronomers postulated that if they came from Biela's comet the comet itself ought to be seen in the southern hemisphere retreating from its encounter with the earth. A German professor telegrammed an observatory in Madras. On November 30th, an astronomer at Madras saw a comet in the place calculated! It was glimpsed again the next night, and then clouds intervened, and not a trace of it was ever seen afterward.

But every year, on the 27th of November, when the earth crosses the orbit of the lost comet, meteoric fragments come plunging into our atmosphere, burning as they fly. Ordinarily their number is small, but when, as in 1872, a swarm of the meteors is in that part of their orbit which the earth crosses, there is a brilliant spectacle. In 1885 this occurred, and the world was treated to one of the most splendid meteoric displays on record. The Biela meteors, also called the Andromedid Meteor Shower, have since faded, though it is possible to glimpse a few each year in late November.

Next let us turn to Perseus. The bending row of stars marking the center of this constellation is very striking and brilliant. The brightest star in the constellation is Alpha, or Mirphak, in the center of the row. The head of Perseus is toward Cassiopeia, and in his left hand he grasps the head of Medusa, which hangs down in such a way that its principal star Beta, or Algol, forms a right angle with Mirphak and Almaach in Andromeda. This star Algol, or the Demon, as the Arabs call it, is in some respects the most wonderful and interesting in all the heavens. It is as famous for the variability of its light as Mira, but it differs widely from that star both in its period, which is very short, and in the extent of the changes it undergoes. During about two days and a half, Algol is equal in brilliance to Mirphak, which is a second-magnitude star; then it begins to fade, and in the course of about four and a half hours it sinks to the fourth magnitude, being then about equal to the faint stars near it. It remains thus obscured for only a few minutes, and then begins to brighten again, and in about four and a half hours more resumes its former brilliance. This phenomenon is very easily observed, for, as will be seen by consulting our little map, Algol can be readily found, and its changes are so rapid that under favorable circumstances it can be seen in the course of a single night to run through the whole gamut.

Of course, no optical instrument whatever is needed to enable one to see these changes of Algol, for it is plainly visible to the naked eye throughout, but it will be found interesting to watch the star with binoculars. Its periodic time from minimum to minimum is two days, twenty hours, and forty-nine minutes, lacking a few seconds. Any one can calculate future minima for himself by adding the periodic time above given to the time of any observed minimum.

The rapid changes of Algol is due to the existence of a faint stellar companion at close quarters in an orbit whose plane is directed edge-wise toward the earth, so that at regular intervals this companion star causes a partial eclipse of Algol.

There was certainly great fitness in the selection of the little group of stars of which this mysterious Algol forms the most conspicuous member, to represent the awful head of the Gorgon carried by the victorious Perseus for the confusion of his enemies. In a darker age than ours the winking of this demon-star must have seemed a prodigy of sinister import.

Turn now to the bright star Mirphak, or Alpha Persei. You will find with your binoculars an exceedingly attractive spectacle there. In my note-book I find this entry, made while sweeping over Perseus for materials for this chapter: *"The field about Alpha is one of the finest in the sky for binoculars. Stars conspicuously ranged in curving lines and streams. A host follows Alpha from the east and south."* The adjacent picture will give the reader some notion of the exceeding beauty of this field of stars, and of the singular manner in which they are grouped, as it were, behind their leader.

The reader will find a starry cluster marked on Map 17 as the "Double Cluster." This object can be easily detected by the naked eye, resembling a wisp of luminous cloud. It marks the hand in which Perseus clasps his diamond sword, and, with a telescope of medium power, it is one of the most marvelously beautiful objects in the sky, a double swarm of stars, bright enough to be clearly distinguished from one another, and yet so numerous as to dazzle the eye with their lively light. Binoculars do not possess sufficient power to resolve this cluster, but they give a startling suggestion of its half-hidden magnificence, and the observer will be likely to turn to it again and again with increasing admiration. Sweep from this to Alpha Persei and beyond to get an idea of the procession of suns in the Milky Way. The nebulous-looking cluster marked M34 appears with binoculars like a faint comet.

About a thousand years ago the theologians undertook to reconstruct the constellation figures, and to give them a religious significance. They divided the zodiac up among the twelve apostles, St. Peter taking the place of Aries, with the Triangles for his mitre. In this reconstruction Perseus was transmogrified into St. Paul, armed with a sword in one hand and a book in the other; Cassiopeia became Mary Magdalene; while poor Andromeda, stripped of all her beauty and romance, was turned into a sepulchre!

Next, look at Cassiopeia, which is distinctly marked out by a zigzag row of stars like a "W" (or "M"). Here the Milky Way is so rich that the observer hardly needs any guidance; he is sure to stumble upon interesting sights for himself. The five brightest stars are generally represented as indicating the outlines of the chair or throne in which the queen sits. Look around Theta (θ) with good binoculars, and you will see a singular and brilliant array of stars near it in a broken half-circle, which may suggest the notion of a crown.

Near the Beta (β) in the map you will see the marking Cas A. This shows the spot where a famous supernova was discovered. It was seen in 1572, and carefully observed by the famous Danish astronomer Tycho Brahe. It seems to have suddenly burst forth with a brilliance that outshone every other star in the heavens, not excepting Sirius itself. But its supremacy was short-lived. In a few months it had sunk to the second magnitude. It continued to grow fainter, exhibiting some remarkable changes of color in the mean time, and in less than a year and a half it disappeared. It has never been seen since, but there remains a strong source of radio emission and a hot expanding bubble of gas. The radio emission comes from a neutron star, the dense remnants of the star that exploded as a supernova.

In 1264, and again in 945, a star is said to have suddenly blazed out near that point in the heavens. There is no certainty about these earlier apparitions, but, assuming that they are not apocryphal, it was thought they might possibly indicate that the star seen by Tycho was a periodical one, its period considerably exceeding three hundred years. Carrying this supposed period back, it was found that an apparition of this star might have occurred about the time of the birth of Christ. It did not require a very prolific imagination to suggest its identity with the so-called Star of Bethlehem and its impending reappearance one day. But this is not the case. The explosion of Tycho's wonderful star was a one-time event, and no one will ever gaze upon its reappearance.

We will pause but briefly with Cepheus, for the old king's constellation is comparatively dim in the heavens, as his part in the dramatic story of Andromeda was contemptible, and he seems to have got among the stars only by virtue of his relationship to more interesting persons. He does possess one gem of singular beauty, the star Mu, which may be found about two and a half degrees south of the star Nu (v). As we learned at the end of the last chapter, Mu is the so-called "Garnet Star," thus named by William Herschel, who advises the observer, in order to appreciate its color, to glance from it to Alpha Cephei, which is a white star. Mu is variable, changing from the fourth to the sixth magnitude in a long period of five or six years. Its color is changeable, like its light. Sometimes it is of a deep garnet hue, and at other times it is orange-colored. Upon the whole, it appears of a deeper red than any other star visible to the naked eye.

Now turn your optics upon the star Delta (δ) Cephei. This is a double star, the components being about forty-one seconds of arc apart, the larger of four and one half magnitude, and the smaller of the seventh magnitude. The latter is of a beautiful blue color, while the larger star is yellow or orange. With a good eye, a steady hand, and a clear glass, magnifying not less than six times, you can separate them, and catch the contrasted tints of their light. Besides being a double star, the brighter component of Delta is an important type of variable star, of which there are many others in our galaxy and other galaxies. You saw one such star, Eta Aquilae, in our tour of the summer stars. These "Cepheid" variables possess the remarkable property that their period of variability is in direct proportion to their instrinsic brightness. Astronomers use this relationship as a cosmic "yardstick" to measure the distances to nearby galaxies. Indeed, it was Edwin Hubble who first measured the period of Cepheid variables in the Andromeda Nebula to determine that it was in fact a galaxy separate from our own Milky Way. This momentous discovery using Cepheid variables led to our understanding of the almost unimaginable distance scale of our universe.

The Stars of Winter

Cold Nights, Bright Stars

I have never watched the first indications of the rising of Orion without a feeling of awakened expectation, like that of one who sees the curtain rise upon a gripping drama. And certainly the magnificent company of the winter constellations, of which Orion is the chief, make their entrance upon the scene in a manner that may be described as dramatic. First in the east come the dazzling Pleiades. At about the same time Capella, one of the most beautiful of stars, is seen flashing above the northeastern horizon. These are the sparkling ushers to the coming spectacle. In an hour the fiery gleam of Aldebaran appears at the edge of the dome below the Pleiades, a star noticeable among a thousand for its color alone, besides being one of the brightest of the heavenly host. The observer familiar with the constellations knows, when he sees this red star which marks the eye of the angry celestial bull, Taurus, that just behind the horizon stands Orion with starry shield and upraised club to meet the charge of his gigantic enemy. With Aldebaran rises the beautiful V-shaped star cluster of the Hyades. Presently the star-streams of Eridanus begin to appear in the east and southeast, the immediate precursors of the rising of Orion:

> *"And now the river-flood's first winding reach The becalmed mariner may see in heaven, As he watches for Orion to espy if he hath aught to say Of the night's measure or the slumbering winds."*

The first glimpse we get of the hero of the sky is the long bending row of little stars that glitter in the lion's skin which, according to mythology, serves him for a shield. The great constellation then advances majestically into sight. First of its principal stars appears Bellatrix in the left shoulder; then the little group forming the head, followed closely by the splendid Betelgeuse, "the martial star," flashing like a decoration upon the hero's right shoulder. Then come into view the equally beautiful Rigel in the left foot, and the striking row of three bright stars forming the Belt. Below these hangs another starry pendant marking the famous sword of Orion, and last of all appears the star Saiph in the right knee. There is no other constellation containing so many bright stars. It has two of the first magnitude, Betelgeuse and Rigel; the three stars in the Belt, and Bellatrix in the left shoulder, are all of the second magnitude; and there are three stars of the third magnitude, more than a dozen of the fourth, and innumerable twinklers of smaller magnitudes, whose scintillations form a celestial illumination of singular splendor.

By the time Orion has chased Taurus the Bull halfway up the eastern slope of the firmament, the peerless Dog-Star, Sirius, is flaming at the edge of the horizon, while farther north glitters Procyon, the little Dog-Star, and still higher are seen the twin stars in Gemini. When these constellations have advanced well toward the meridian overhead, as shown in our circular map, their united radiance forms a scene never to be forgotten. Counting one of the stars in Gemini as of the first rank, there are no less than seven first-magnitude stars ranged around one another in a way that can not fail to attract the attention and the admiration of the most careless observer. Aldebaran, Capella, the Twins Castor and Pollux, Procyon, Sirius, and Rigel mark the angles of a huge hexagon, while Betelgeuse shines with ruddy beauty not far from the center of the figure. The heavens contain no other naked-eye view comparable with this great array, not even the glorious celestial region where the Southern Cross shines supreme, being equal to it in splendor.

As an offset to the discomforts of winter observations of the stars, you will find that the softer skies of summer have no such marvelous brilliants to dazzle your eyes as those that light the

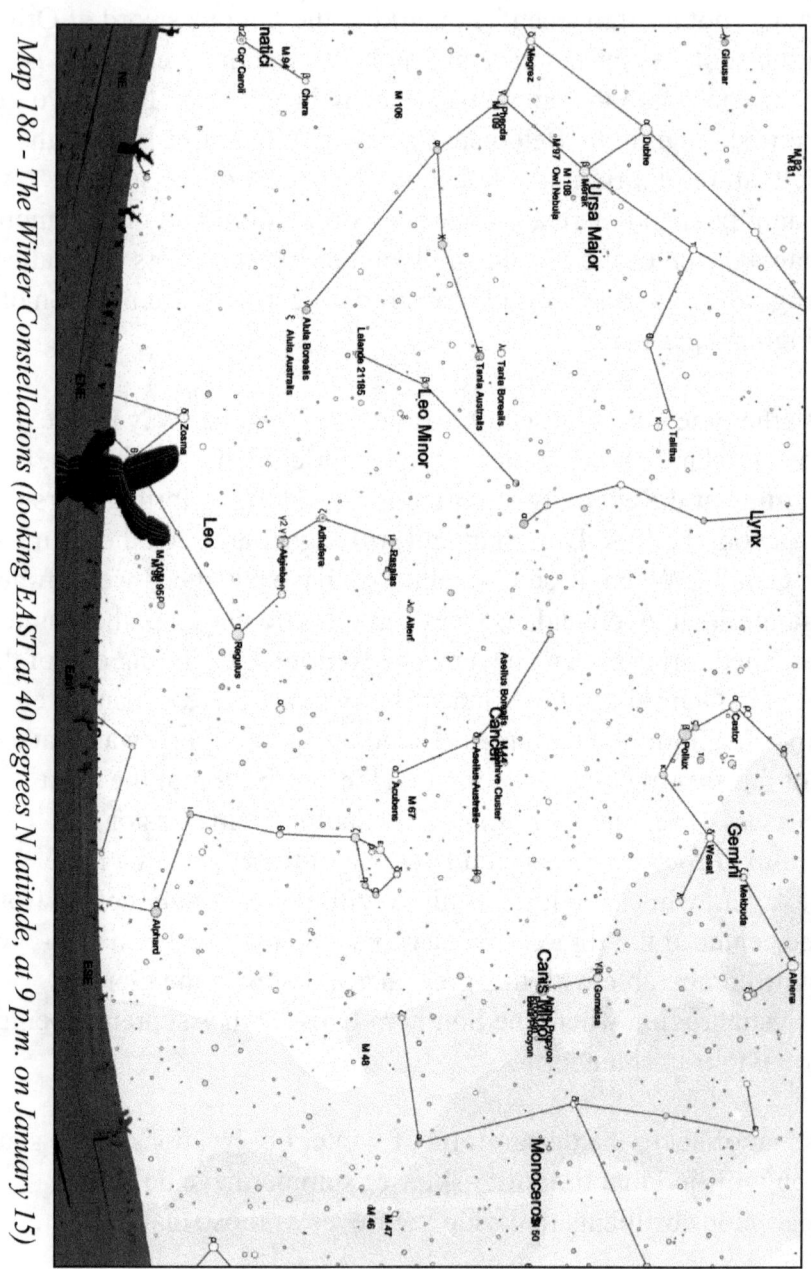

Map 18a - The Winter Constellations (looking EAST at 40 degrees N latitude, at 9 p.m. on January 15)

114

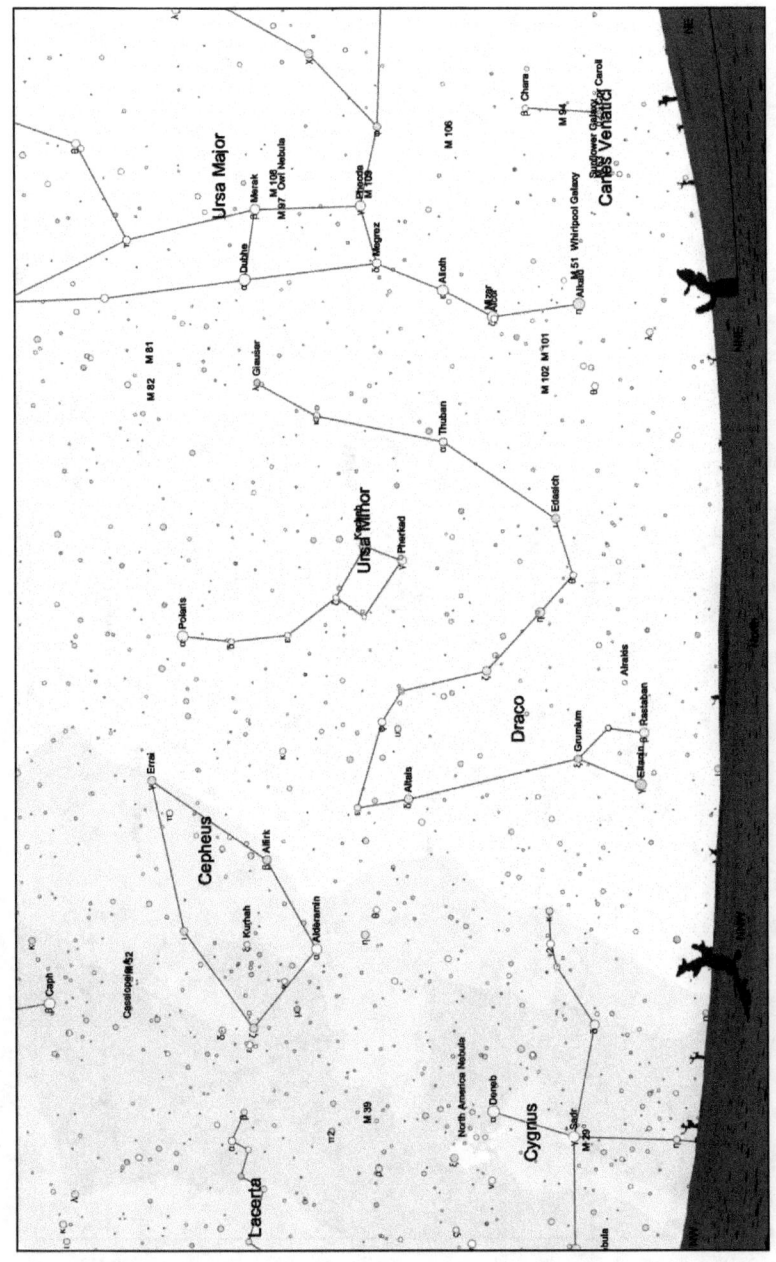

Map 18b - The Winter Constellations (looking NORTH at 40 degrees N latitude, at 9 p.m. on January 15)

115

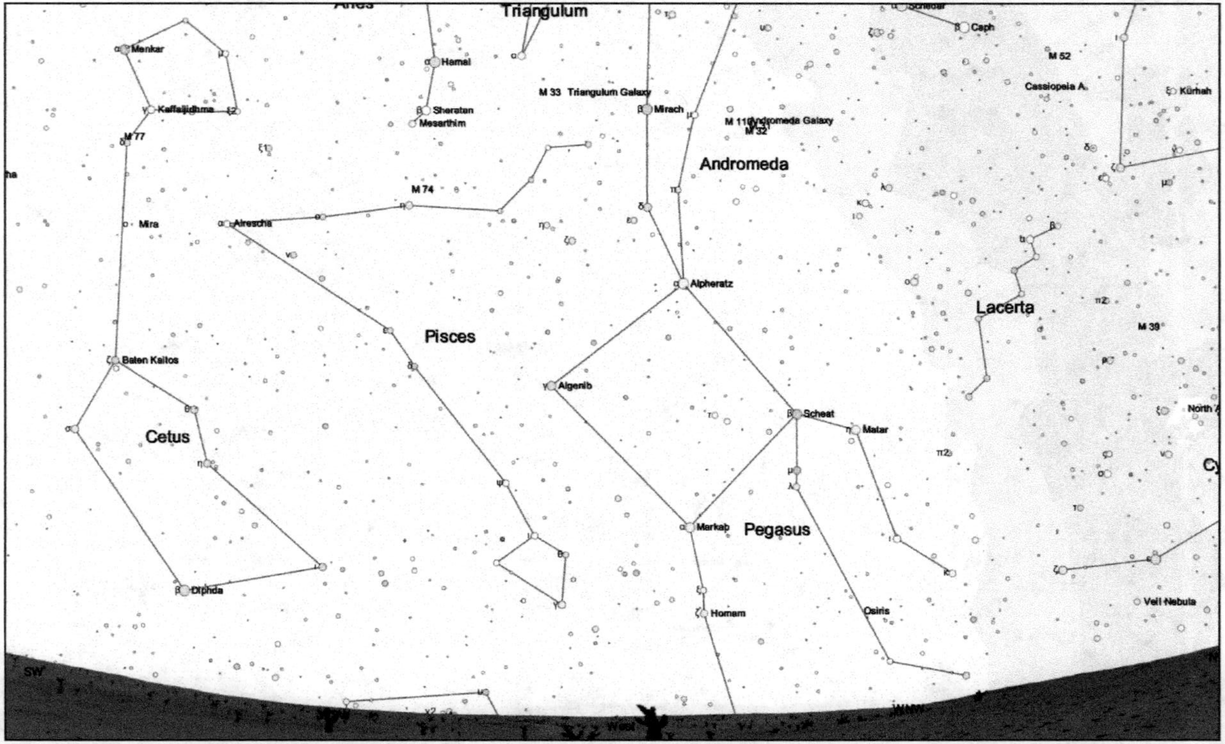

Map 18c - The Winter Constellations (looking WEST at 40 degrees N latitude, at 9 p.m. on January 15)

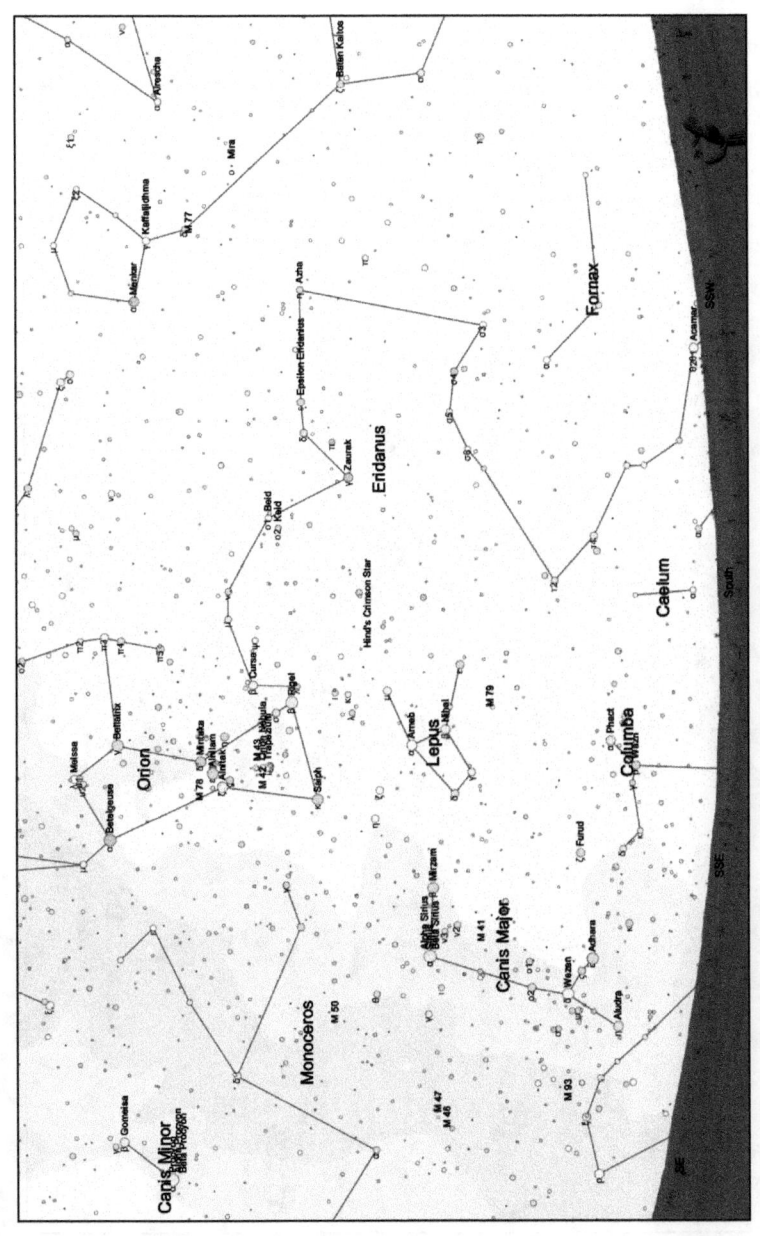

Map 18d - The Winter Constellations (looking SOUTH at 40 degrees N latitude, at 9 p.m. on January 15)

117

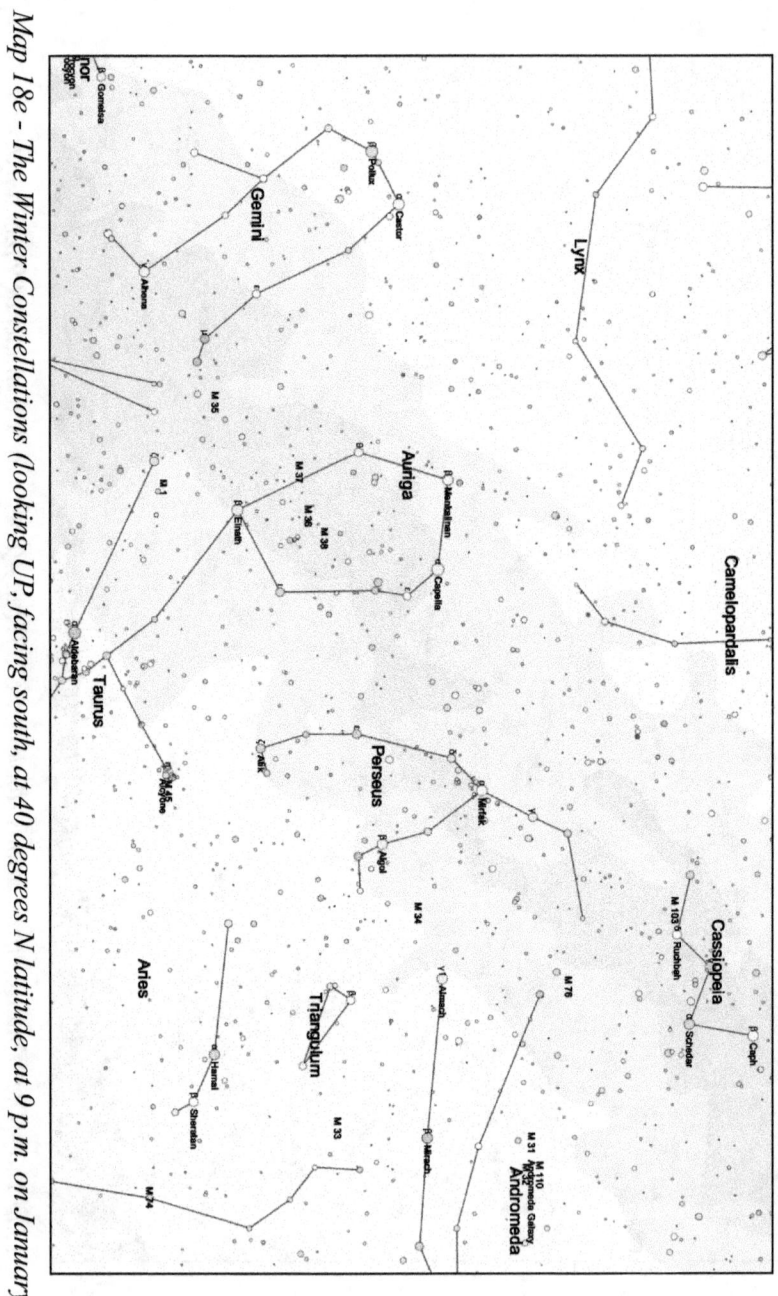

Map 18e - The Winter Constellations (looking UP, facing south, at 40 degrees N latitude, at 9 p.m. on January 15)

winter heavens. To comprehend the real glories of the celestial sphere in the depth of winter, one should spend a few clear nights in the countryside of northern latitudes, when the hills, clad with sparkling blankets of crusted snow, reflect the glitter of the evening sky. In the pure frosty air the stars seem splintered and multiplied indefinitely, and the brighter ones shine with a splendor of light and color unknown to the denizen of the smoky city, whose eyes are dulled and blinded by the glare of streetlights and skyscrapers. There, one may detect the delicate shades that lurk in the imperial

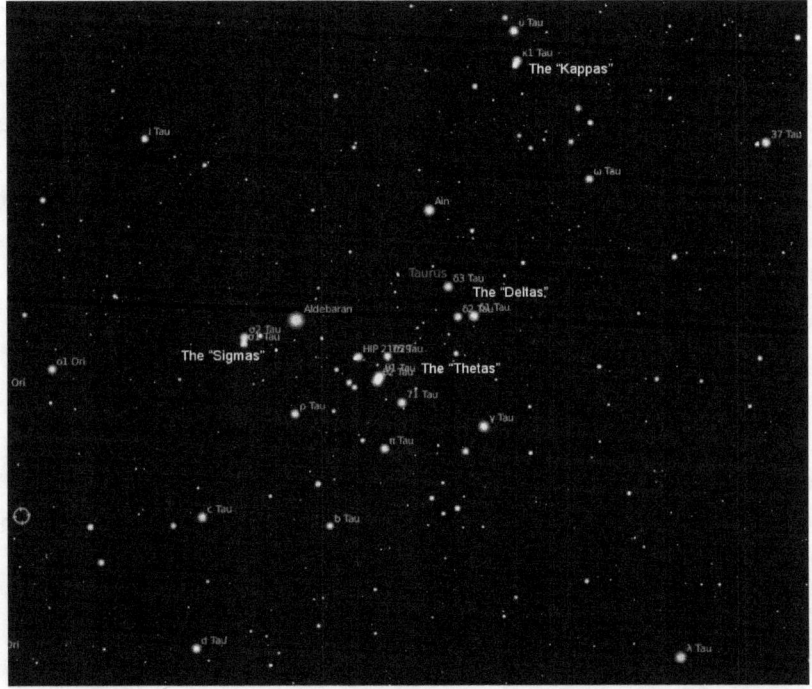

The Hyades

blaze of Sirius, the beautiful rose-red light of Aldebaran, the rich orange hue of Betelguese, the blue-white radiance of Rigel, and the pearly luster of Capella. If you have never seen the starry heavens except as they appear from city streets and squares, then, I had almost said, you have never seen them at all, and especially in the

119

winter is this true. I wish I could describe to you the impression that they can make upon the opening mind of a country boy, who, knowing as yet nothing of the little great world around him, stands in the yawning silence of night and beholds the great world above him, looking deep into the shining vistas of the universe, and overwhelmed with the wonder of those glittering stars.

Looking now at Map 18, we see the heavens as they appear at midnight on the 1st of December, at 10 o'clock p. m. on the 1st of January, and at 8 o'clock p.m. on the 1st of February. In the western half of the sky we recognize Andromeda, Pegasus, Pisces, Cetus, Aries, Cassiopeia, and other constellations that we studied in the "Stars of Autumn." Far over in the east we see rising Leo, Cancer, and Hydra, which we included among the "Stars of Spring." Occupying most of the southern and eastern heavens are the constellations which we are now to describe under the name of the " Stars of Winter," because in that season they are seen under the most favorable circumstances. I have already referred to the admirable way in which the principal stars of some of these constellations are ranged around one another.

By the aid of the map you can perceive the relative position of the different constellations, and, having fixed this in your mind, you will be prepared to study them in detail. Let us now begin with Map No. 19, which shows us the constellations of Eridanus, Lepus, Orion, and Taurus.

Eridanus, The River

Eridanus is a large though not very conspicuous constellation, which is generally supposed to represent the celebrated river now known as the Po. It has had different names among different peoples, but the idea of a river, suggested by its long, winding streams of stars, has always been preserved. According to fable, it is the river into which Phaeton fell after his disastrous attempt to drive the chariot of the sun for his father Phoebus, and in which hare-brained adventure he narrowly missed burning the world up. The imaginary river starts from the brilliant star Rigel, in the left foot of Orion, and flows in a broad upward bend toward the west; then it turns in a southerly direction until it reaches the bright star Gamma (γ), where it bends sharply to the north, and then quickly sweeps off to the west once more, until it meets the group of stars marking the head of Cetus. Then it runs south, gradually turning eastward, until it flows back more than half-way to Orion. Finally it curves south again and disappears beneath the horizon. Throughout the whole distance of more than 100 degrees, the course of the stream is marked by rows of stars, and can be recognized without difficulty by the casual observer.

The first thing to do with your binoculars, after you have fixed the general outlines of the constellation in your mind by naked-eye observations, is to sweep slowly over the whole course of the stream, beginning at Rigel, and following its various wanderings. Eridanus ends in the southern hemisphere near a first-magnitude star called Achernar, which is situated in the stream, but can not be seen from northern latitudes. Along the stream you will find many interesting groupings of the stars. In the map see the pair of stars below and to the right of Nu (ν). These are the two Omicrons, the upper one being Omicron 1 and the lower one Omicron 2. The latter is of an orange hue, and is remarkable for the speed with which it appears to move across the sky. There are only one or two stars whose proper motion, as it is called, is more rapid than that of Omicron 2 in Eridanus. It changes its place nearly seven minutes of arc in a century. The records of the earliest observations we possess show that near the beginning of the Christian era it was about half-way between Omicron 1 and Nu. Its companion Omicron 1 on the contrary, seems to be almost stationary, so that Omicron 2 will gradually draw away from it, passing on toward the southwest until, in the course of centuries, it will become invisible from our latitudes. This flying star is accompanied by two faint companions, which in themselves form a close and very delicate double star. These two little stars, of only 9.5 and 10.5 magnitude, respectively, are, of course beyond the sight of the observer with modest binoculars. The system of which they form a part, however, is intensely interesting, since the appearances indicate that they belong, in the manner of satellites, to Omicron 2, and are fellow voyagers of that wonderful star. Omicron 2 is also catalogued as 40 Eridani. The system lies just 16.5 light years from Earth.

Taurus, The Bull

Having admired the star-groups of Eridanus, one of the prettiest of which is to be seen around Beta (β), also called Cursa, let us turn next to Taurus, just above or north of Eridanus. Two remarkable clusters at once attract the eye, the Hyades, which are shaped somewhat like the letter V, with Aldebaran in the upper end of the left-hand branch, and the Pleiades, whose silvery glittering has made them celebrated in all ages. The Pleiades are in the shoulder and the Hyades in the face of Taurus, Aldebaran most appropriately representing one of his blazing eyes as he hurls himself against Orion. The constellation makers did not trouble themselves to make a complete Bull, and only the head and forequarters of the animal are represented. If Taurus had been completed on the scale on which he was begun, there would have been no room in the sky for Aries; one of the Fishes would have had to abandon his celestial swimming place, and even the fair Andromeda would have found herself uncomfortably situated. But, as if to make amends for neglecting to furnish their heavenly Bull with hind-quarters, the ancients gave him a most prodigious and beautiful pair of horns, which make the beholder feel alarm for the safety of Orion.

Starting out of the head above the Hyades, as illustrated in Map 19, the horns curve upward and to the east, each being tipped by a bright star. Along and between the horns runs a scattered and broken stream of minute stars which seem to be gathered into knots just beyond the end of the horns, where they dip into the edge of the Milky-Way. Many of these stars can be seen, on a dark night, with ordinary binoculars. With such a glass their appearance almost makes one suspect that Virgil had a poetic prevision of the wonders yet to be revealed by the telescope when he wrote of the season:

"When with his golden horns in full career
The Bull beats down the barriers of the year."

Below the tips of the horns, and over Orion's head, there are also rich clusters of stars, as if the Bull were flaunting shreds of sparkling raiment torn from some celestial victim of his fury. With an ordinary glass, however, the observer will not find in this star-sprinkled region around the horns of Taurus as brilliant a spectacle as that presented by the Hyades and the group of stars just above them in the Bull's ear.

The two stars in the tips of the horns are both interesting, each in a different way. The upper and brighter one of the two, not shown in Map 19, is called Alnath. It is common to the left horn of Taurus and the right foot of Auriga, who is represented standing just above. It is a singularly white star. This quality of its light becomes conspicuous when it is looked at with a glass. The most inexperienced observer will hardly fail to be impressed by the pure whiteness of Alnath, in comparison with which he will find that many of the stars he had supposed to be white show a decided tinge of color.

The star in the tip of the right or southern horn, Zeta (ζ), is remarkable, not on its own account, but because it serves as a pointer to a famous nebula, the discovery of which led Messier to form his catalogue of nebulae. This is sometimes called the "Crab Nebula", from the long sprays of nebulous matter which were seen surrounding it with Lord Rosse's great telescope. Our map is simply intended to enable the observer to locate this strange object. If he wishes to study its appearance, he must use a powerful telescope. But with good binoculars in dark sky, he can see it as a speck of light in the position shown in the Map, where the large star is Zeta and the smaller ones are faint stars, the relative position of which will enable the observer to find the nebula, if he keeps in mind that the top of the Map is toward the north. It is noteworthy that this nebula for a time deceived several of the watchers who were on the lookout for the predicted return of Halley's comet in 1835 and again in 1986..

The Crab Nebula is the remnant of a supernova explosion which occurred in 1054 A.D. Like Tycho's star in Cassiopeia, which we met in the last chapter, this star was view by ancient astronomers in China and Arabia. It was visible for some months before it faded from view, which is common for this type of dying star.

And now let us look at the Hyades, an assemblage of stars not less beautiful than their more celebrated sisters the Pleiades. The leader of the Hyades is Aldebaran, or Alpha Tauri, and his followers are worthy of their leader. The inexperienced observer is certain to be surprised by the display of stars which binoculars bring to view in the Hyades. Our illustration will give some notion of their appearance. A "brackish poet," of whose rhymes the amateur astronomer Admiral Smyth was so fond, thus describes the Hyades:

" In lustrous dignity aloft see Alpha Tauri shine, The splendid zone he decorates attests the Power divine:
For mark around what glitt'ring orbs attract the wandering eye, You'll soon confess no other star has such attendants nigh."

The redness of the light of Aldebaran is a very interesting phenomenon. Careful observation detects a decided difference between its color and that of Betelgeuse, or Alpha Orionis, which is also a red star. It differs, too, from the brilliant red star of summer, Antares. Aldebaran has a trace of rose-color in its light,

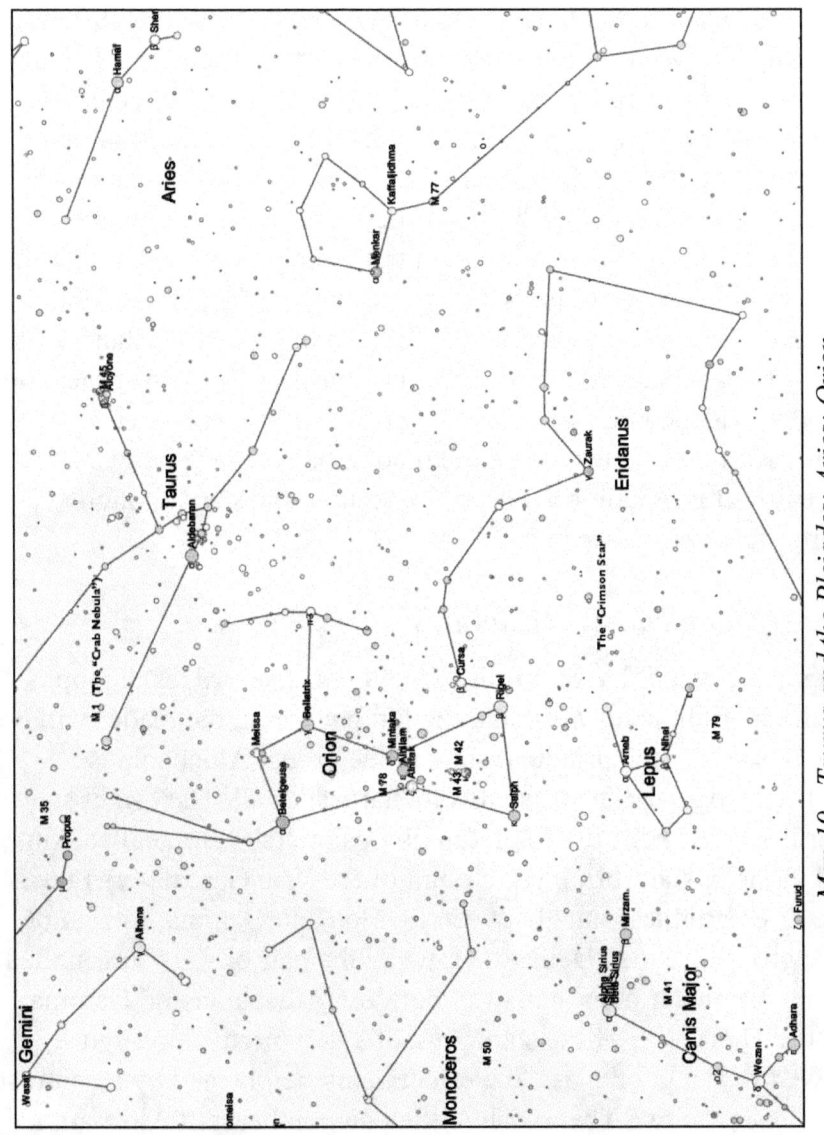

Map 19 - Taurus and the Pleiades; Aries; Orion

125

while Betelgeuse is of a very deep orange, and Antares may be described as fire-red. These shades of color can easily be detected by the naked eye after a little practice. First compare Aldebaran and Betelgeuse, and glance from each to the brilliant bluish-white, star Rigel in Orion's foot. Upon turning the eye back from Rigel to Aldebaran the peculiar color of the latter is readily perceived. Spectroscopic analysis has revealed the presence in Aldebaran of hydrogen, sodium, magnesium, calcium, iron, bismuth, tellurium, antimony, and mercury. The star is beyond its middle age and slowing swelling into a red giant, after which it will end its life in by expelling its outer layers as a planetary nebula. And so modern discoveries, while they have pushed back the stars to distances of which the ancients could not conceive, have, at the same time, and equally, widened our understanding of the physical universe and abolished forever the ancient distinction between the heavens and the earth. It is a plain road from the earth to the stars, though mortal feet can not tread it.

The Hyades and The Pleiades

Keeping in mind that in our little picture of the Hyades the top is north, the right hand west, and the left hand east, the reader will be able to identify the principal stars in the group. Aldebaran is readily recognized, because it is the largest of all. The bright star near the upper edge of the picture is Epsilon (ε) Tauri, called Ain, and its sister star, forming the point of the V, is Gamma (γ) Tauri. The three brightest stars between Epsilon and Gamma, forming a little group, are the "Deltas" (δ), while the pair of stars surrounded by many smaller ones, half-way between Aldebaran and Gamma, are the Thetas (θ). These stars present a very pretty appearance, viewed with a good glass, the effect being heightened by a contrast of color in the two Thetas. The little pair southeast of Aldebaran, called the Sigmas, is also a beautiful sight. The distance apart of these stars is about seven minutes of arc, while the distance between the two Thetas is about five and a half minutes of arc.

These measures may be useful to the reader in estimating the distances between other stars that he may observe. It will also be found an interesting test of the eye-sight to endeavor to see these stars as doubles without the aid of a glass. Persons having keen eyes will be able to accomplish this.

North of the star Epsilon will be seen a little group in the ear of the Bull, which presents a brilliant appearance with a small glass. The southernmost pair in the group are the "Kappas" (κ), whose distance apart is very nearly the same as that of the Thetas, described above; but I think it improbable that anybody could separate them with the naked eye, as there is a full magnitude between them in brightness, and the smaller star is only of magnitude 6.5, while sixth-magnitude stars are generally reckoned as the smallest that can be seen with the naked eye. Above the Kappas, and in the same group in the ear, are the two Upsilons, forming a wider pair.

Next we come to the Pleiades:

"Though small their size and pale their light, wide is their fame."

In every age and in every country the Pleiades have been watched, admired, and wondered at, for they are visible from every inhabited land on the globe. To many they are popularly known as the Seven Sisters, although few persons can see more than six stars in the group with the unaided eye. It is a singular fact that many of the earliest writers declare that only six Pleiades can be seen, although they all assert that they are seven in number. These seven were the fabled daughters of Atlas, or the Atlantides, whose names were Merope, Alcyone, Celaeno, Electra, Taygeta, Asterope, and Maia. One of the stories connected with them is that Merope married a mortal, whereupon her star grew dim among her sisters. Another fable assures us that Electra, unable to endure the sight of the burning of Troy, hid her face in her hands, and so blotted her star from the sky. While we may smile at these stories, we can not entirely disregard them, for they are intermingled with some of the richest literary treasures of the world, and they come to us, like some old keepsake, perfumed with the memory of a past age.

The mythological history of the Pleiades is intensely interesting, too, because it is world-wide. They have impressed their mark, in one way or another, upon the habits, customs, tradition, language, and history of probably every nation. This is true of simple tribes and great empires. The Pleiades furnish one of the principal links that appear to connect the beginnings of human history with that wonderful prehistoric past, where, as through a gulf of mist, we seem to perceive the glow of a golden age beyond. The connection of the Pleiades with traditions of the Flood is most remarkable. In almost every part of the world, and in various ages, the celebration of a feast or festival of the dead, dimly connected by traditions with some great calamity to the human race in the past, has been found to be directly related to the Pleiades. This festival or rite, which has been discovered in various forms among the ancient Hindu, Egyptians, Persians, Peruvians, Mexicans, and Druids, occurs always in the month of November, and is regulated by the culmination of the Pleiades. The Egyptians directly connected this celebration with a deluge, and the Mexicans, at the time of the Spanish conquest, had a tradition that the world had once been destroyed at the time of the midnight culmination of the Pleiades. Among the original inhabitants of Australia and the Pacific islands a similar rite has been discovered. It has also been suggested that the Japanese, who called the cluster *Subaru*, associated the feast of lanterns with this worldwide observance of the Pleiades, as commemorating some calamitous event in the far past which involved the whole race of man in its effects.

The Pleiades also have a supposed connection with that mystery of mysteries, the great Pyramid of Cheops. It has been found that about the year 2170 B.C., when the beginning of spring coincided with the culmination of the Pleiades at midnight, that wonderful group of stars was visible, just at midnight, through the mysterious southward-pointing passage in the Pyramid. At the same date the pole-star, which in that distant age was not Polaris by Alpha Draconis, was visible through the northward-pointing passage of the Pyramid.

Another curious myth involving the Pleiades as a part of the constellation Taurus is that which represents this constellation as the Bull into which Jupiter (or Zeus) changed himself when he carried the fair Europa away from Phoenicia to the continent that now bears her name. In this story the fact that only the head and fore-quarters of the Bull are visible in the sky is accounted for on the ground that the remainder of his body is beneath the water through which he is swimming. Here, then, is another apparent link with the legends of the Flood, with which the Pleiades have been so strangely connected, as by common consent among many nations, and in the most widely separated parts of the earth.

With binoculars, you may be able to see all of the stars represented in our picture of the Pleiades. The scene is a remarkable one, perhaps the finest for binoculars in the entire heavens. Not only all of the "Seven Sisters," but many other stars, can be seen twinkling among them. The superiority of Alcyone to the others, which is not so clear to the naked eye, becomes very apparent. Alcyone is the large star left of the middle of the picture with a triangle of little stars beside it. To the left or east of Alcyone the two most conspicuous stars are Atlas and Pleione. It requires a sharp eye to see Pleione without a glass, while Atlas is plainly visible to the unaided vision, and is always counted among the naked-eye Pleiades,

The Pleiades, with associated nebulosity.

although it does not bear the name of one of the mythological sisters, but that of their father. The bright star below and to the right of Alcyone is Merope; the one near the right-hand edge of the picture, about on a level with Alcyone, is Electra. Above, or to the north of Electra, are two bright stars lying in a line pointing toward Alcyone ; the upper one of these, or the one farthest from Alcyone, is Taygeta, and the other is Maia. Above Taygeta and Maia, and forming a little triangle with them, is a pair of stars which bears the name of Asterope. About half-way between Taygeta and Electra, and directly above the latter, is Celaeno.

The naked-eye observer will probably find it difficult to decide which he can detect the more easily, Celaeno or Pleione, while he will discover that Asterope, although composed of two stars, as seen with a glass, is so faint as to be much more difficult than either Celaeno or Pleione. Unless, as is not improbable, the names have become interchanged in the course of centuries, the brightness of these stars would seem to have undergone remarkable changes. The star of Merope, it will be remembered, was said to have become indistinct, or disappeared, because she married a

130

mortal. At present Merope is one of those that can be plainly seen with the naked-eye, while the star of Asterope, who was said to have had the god Mars for her spouse, has faded away until only a glass can show it. It would appear, then, that notwithstanding an occasional temporary eclipse, it is, in the long run, better to marry a plain mortal than a god. Electra, too, who hid her eyes at the sight of burning Troy, seems to have recovered from her fright, and is at present, next to Alcyone, the brightest star in the cluster. But, however we may regard those changes in the brightness of the Pleiades which are based upon tradition, it is possible that changes may have taken place in the comparative brilliancy of stars in this cluster since astronomy became an exact science.

Observations of the proper motions of the Pleiades have shown that there is an actual physical connection between them; that they are, literally speaking, a flight of suns. Their common motion is toward the southwest. The stars are bound to each other gravitationally, having formed out of a common cloud of gas and dust some 100 million years ago. As they are perturbed by passing stars, the sisters will eventually disperse into the plane of the galaxy and evolve as lone stars.

In 1859, an extensive blue-white nebula was discovered, of a broad oval form, with the star Merope immersed in one end of it. It was thought the nebula is caused by starlight reflecting off residual gas and dust left over from the formation of the cluster. But modern studies suggest the dust is simply a part of the interstellar medium through which the cluster is passing.

Although the nebulosity is easily seen on time-exposure photographs, the reader should not expect to be able to see the nebula in the Pleiades with anything other than binoculars of 70-80 mm or more in aperture and pristine dark sky. Still, this star cluster, like the Hyades, is one of few that looks finer in binoculars than a telescope. You are encouraged to return to it again and again. It will provide you, as it did Tennyson, with countless nights of pleasant viewing:

"Many a night I saw the Pleiades, rising through the mellow shade.
Glitter like a swarm of fireflies tangled in a silver braid."

Orion, The Hunter

Orion will next command our attention. You will find the constellation in Map 19:

"Eastward beyond the region of the Bull
Stands great Orion; who so kens not him in cloudless night,
Gleaming aloft, shall cast his eyes in vain
To find a brighter sign in all the heaven."

To the naked eye, to binoculars, and to the telescope, Orion is a gold mine of wonders. This great constellation embraces almost every variety of interesting phenomena that the heavens contain. Here we have the grandest of the nebulae, some of the largest and most beautifully colored stars, star-streams, star-clusters, nebulous stars, variable stars. I have already mentioned the positions of the principal stars in the imaginary figure of the great hunter. I may add that his upraised arm and club are represented by the stars seen in the map above Alpha (α) or Betelgeuse, one of which is marked Nu (ν), and another, in the knob of the club, Chi (χ). I have also, in speaking of Aldebaran, described the contrast in the colors of Betelgeuse and Beta (β) or Rigel. Betelgeuse, it may be remarked, is slightly variable. Sometimes it appears brighter than Rigel, and sometimes less brilliant. Betelgeuse is a cool but massive red supergiant star, possessed of an absorptive envelope or shell, and which will one day end its life in a spectacular supernova explosion that will light the day and nighttime skies of Earth for many weeks. Rigel, on the other hand, is still in its prime of life; it's hotter and younger than Betelguese. So, then, we may look upon the two chief stars of this great constellation as representing two stages of stellar existence. Betelgeuse shows us a sun that has almost run its course, that has passed into its decline before the on-coming and inevitable fate of extinction; but in Rigel we see a sun blazing with the fires of youth, splendid in the first glow of its solar energies, and holding the promise of the future yet before it. We may pursue this comparison one step farther back in time and see in the famed Great Nebula of Orion, M42, which glows dimly in the middle of the constellation, between Rigel and Betelgeuse. The nebula represents a still earlier cosmical condition: the birth of stars whose infant rays will illuminate space when Rigel itself is growing dim and Betelguese is long passed.

Turn your glass upon the three stars forming the Belt. You will not be likely to undertake to count all the twinkling lights that you will see, especially as many of them appear and disappear as you turn your attention to different parts of the field. The stars of the belt are, from east to west, Alnitak, Alnilam, and Mintaka. Though only visible in photographs, Alnitak is embedded in a rich nebulosity overlaid with black dust in the shape of a horse's head, giving the nebula in this region the name "The Horse Head Nebula." As you sweep around the Belt of Orion, notice the sweeping "S-shaped" group of stars that begins just north of Mintaka, sweeps down between Mintaka and Alnilam, ending below the latter.

Sweep all around the Belt and also between the Belt and Gamma (γ) or Bellatrix. According to the old astrologers, women born under the influence of the star Bellatrix were considered lucky.

Below the Belt will be seen a short row of stars hanging downward and representing the sword. In the middle of this row is the Great Orion Nebula, which lies some 1200 light years from our planet. Just above Iota (ι), you will find the star Theta (θ) embedded in the nebula. This star has two components easily resolved in binoculars. Their position suggests, correctly, that these hot blue stars recently formed out of the diffuse gas and dust of the nebula, and are in fact lighting the nebula and setting it aglow from within. One of the components of Theta is a quadruple star called the Trapezium. A small telescope is required to resolve these four fledgling stars.

Hydrogen is abundant in this regions of space, but the constituent parts of this elemental gas, the proton and neutron, are ripped apart by the energetic blue and ultraviolet light from newly-lit stars. The glow of the Orion Nebula is caused by the recombination of electrons with protons to re-form hydrogen. As the charged particles come together, they release energy as red and green light. There are many such nebula in the sky, but this is the most spectacular for visual observers, with the exception of the Eta Carina nebula in the deep southern sky.

Binoculars reveal much of the extent of the nebula you see in photographs. Avert your eyes slightly to the expose the most sensitive part of your retina to the light. Even in modest binoculars, you will see the great extent of the nebula as it is seen in photographs, extending over a span of sky twice the diameter of the full moon. You will not, however, see color with binoculars or small telescopes because the light is too feeble to stimulate the color receptors of your eye.

Millions of years hence, this delicate nebula will coalesce into an open star cluster that will serve as a fine telescopic objects for our distant descendants. As an example of such a star cluster, you will observe just north of M42 an open star cluster called NGC 1981, a scattering of ten blue-white stars. It makes a pretty contrast adjacent to the nebula of fire-mist within the same field of view.

Other stars are seen scattered in different parts of the Sword. Iota has a lovely silver-blue color, and is one of dozens of blazing hot stars in Orion. Only 8 arcminutes south of Iota is a striking double star called Struve 747, named after the famed astronomer who catalogued hundreds of such stars. The comparable brightness and generous separation make this star easily resolved and one of the best in the sky for binoculars.

Modern measurements reveal the entire region of Orion is embedded in cold dust and gas that will one day yield new stars. The Great Nebula itself is but a tiny transitory blister of star formation. Many more such regions have coalesced in the past, and many more will coalesce over the coming millions of years.

The Great Nebula in Orion

Do not fail to look for two little stars just west of Rigel, which, with modest binoculars, appear to be almost hidden in the flashing rays of its brilliant companion.

With a telescope, Rigel is one of the most beautiful double stars in the sky, having a little blue companion close under its wing.

Run your glass along the line of little stars forming the lion's skin or shield that Orion opposes to the onset of Taurus. Here you will find some interesting combinations, and the star marked on the map π^5 will especially attract your eye, because it is accompanied, about fifteen minutes to the northwest, by a seventh-magnitude star of a rich orange hue.

Look next at the little group of three stars forming the head of Orion. Although there is no nebula here, these stars, as seen with the naked eye, have a remarkably nebulous look, and Ptolemy regarded the group as a nebulous star. The largest star is called Lambda (λ) ; the others are Phi (φ) 1 and 2. Binoculars will show another star above Lambda, and a fifth star below Phi 2 , which is the farthest of the two Phis from Lambda. It will also reveal a faint twinkling between Lambda and Phi 2. Binoculars show this twinkling is produced by a pretty little row of three stars of the eighth and ninth magnitudes.

In fact, Orion is such a striking object in the sky that more than one attempt has been made to steal away its name and substitute that of some modern hero. The University of Leipsig, in 1807, formally resolved that the stars forming the Belt and Sword of Orion should henceforth be known as the constellation of Napoleon. As if to offset this, an Englishman proposed to rename Orion for the British naval commander Nelson. But *"Orion armed"* has successfully maintained his name and place against all comers. As becomes the splendor of his constellation, Orion is a tremendous hero of antiquity, although it must be confessed that his history is somewhat shadowy and uncertain, even for a mythological story. All accounts agree, however, that he was the mightiest hunter ever known, and the Hebrews claimed that he was no less a person than Nimrod himself.

The little constellations of Lepus and Columba, below Orion, need not detain us long. You will find in them some pretty combinations of stars. In Lepus is the celebrated "Crimson Star," which has been described as resembling a drop of blood in a truly marvelous hue for a sun. You will find it on a line extending from Alpha Lepus (Arneb) through Mu (μ), at an distance west of Mu of just 3 degrees.

Canis Major; Sirius; Puppis

We will now turn to the constellation of Canis Major, represented in Map 20. Although, as a constellation, it is not to be compared with the brilliant Orion, yet, on account of the unrivaled magnificence of its chief star, Canis Major presents almost as attractive a scene as its more extensive rival. Everybody has heard of Sirius, or the Dog Star, and everybody must have seen it flashing and scintillating so splendidly in the winter heavens. Sirius, in fact, stands in a class by itself as the brightest star in the sky. Its light is white, with a shade of blue, which requires close watching to be detected. When it is near the horizon, or when the atmosphere is very unsteady, Sirius flashes prismatic colors like a great diamond. The question has been much discussed, as to whether Sirius was formerly a red star. It is described as red by several ancient authors, but it seems to be pretty well established that most of these descriptions are due to a blunder made by Cicero in his translation of the astronomical poem of Aratus. It is not impossible, though it is highly improbable, that Sirius has changed color.

So intimately was Sirius connected in the minds of the ancient Egyptians with the annual rising of the Nile, that it was called the Nile-star. When it appeared in the morning sky, just before sunrise, the season of the overflowing of the great river was about to begin, and so the appearance of this star was regarded as foretelling the coming of the floods. The "dog days" got their name from Sirius, as they occur at the time when that star rises with the sun.

Your eyes will be fairly dazzled when you turn your binoculars upon this splendid star. By close attention you will perceive a number of faint stars, mere points by comparison, in the immediate neighborhood of Sirius. There are many interesting objects in the constellation. The star marked Nu (v2) in the map, just south and west of Sirius, is really triple, as the smallest glass will show. Look next at the star cluster M41 below and to the east of Sirius. The

137

cloud of minute stars of which it is composed can be very well seen with binoculars, and presents a splendid view in a small telescope. The star Sigma (σ) is of a very ruddy color that contrasts beautifully with the light of Epsilon (ε), which can be seen in the same field of view with binoculars. Between the stars Delta (δ), also called Wezen, and Omicron (o) 1 and Omicron 2 there is a remarkable array of minute stars. The open star cluster NGC 2362 lies just northeast of Wezen and appears in binoculars as a frosty patch of blue-white stars, with the star Tau set in its midst.

Sirius owes its extraordinary brilliancy to its intrinsic brightness and to its closeness to Earth. It is 8.6 light years away, making it the fifth nearest star to our own sun. It is more massive than our sun by 2 times, and shines some 25 times brighter. While this seems impressive, the intrinsic brightness of Sirius is greatly exceeded by such massive stars as Rigel, Deneb, and Canopus, which are outshone by Sirius only as a consequence of their much greater distance.

Sirius, as we saw when talking of Procyon (see Chapter I), is a double star. For many years after Bessel had declared his belief that the Dog-Star was subjected to the attraction of an invisible companion, telescopes failed to reveal the accompanying star. Finally, in 1862, a new telescope that Alvan Clark had just finished and was testing, brought the hidden star into view. The star is exceedingly faint yet possesses a mass similar to that of our sun. As we now know, and mentioned in Chapter I, the faint but massive companion stars to Sirius and Procyon are the dense remains of a mid-sized star that has settled into its final phase of life as a white dwarf.

The great writer and philosopher Voltaire penned an extraordinary short story called "Micromegas", in which the hero came from an imaginary planet circling around Sirius. Inasmuch as Voltaire, together with Jonathan Swift who ascribed two moons to Mars

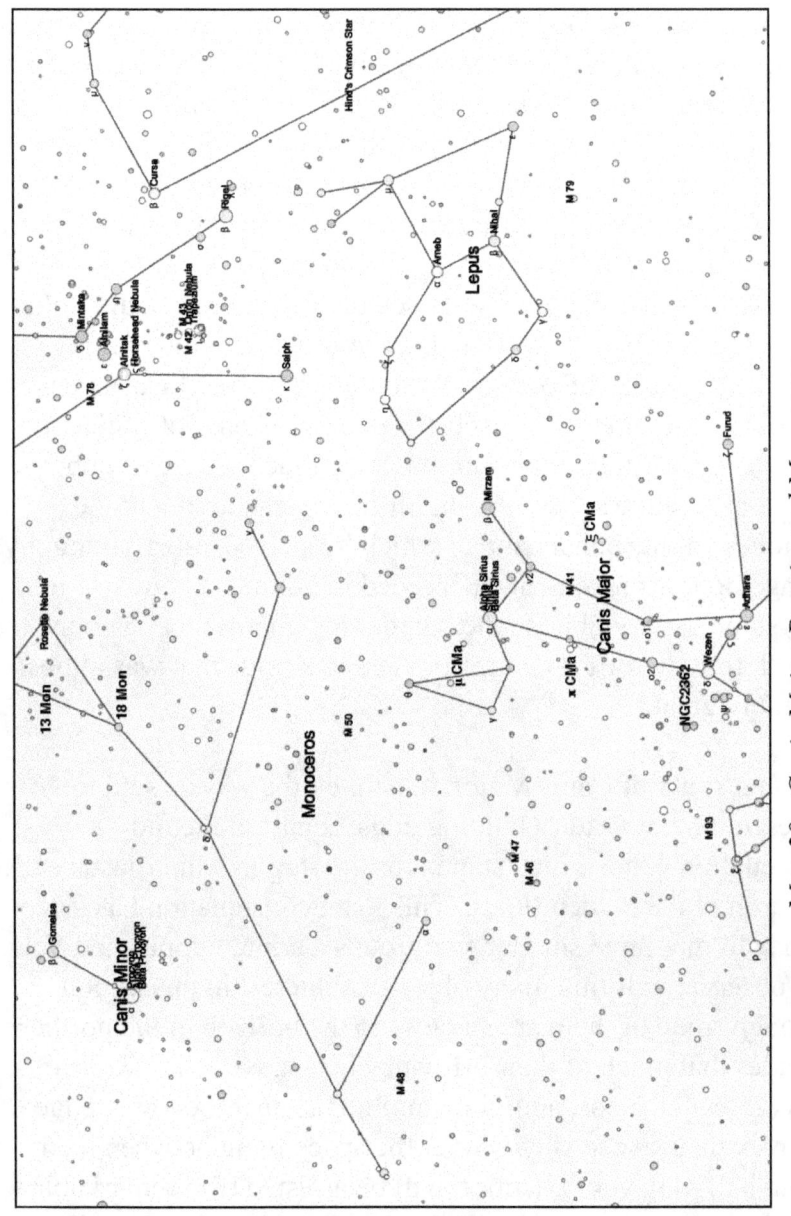

Map 20 - Canis Major, Puppis, and Monoceros

many years before they were discovered, it is all the more interesting that Voltaire should have imagined an enormous planet circling around the Dog-Star. But Voltaire went far astray when he ascribed a gigantic stature to his "Sirian." He makes Micromegas, whose world was 21,600,000 times larger in circumference than the earth, more than twenty miles tall, so that when he visited our little planet he was able to wade through the oceans and step over the mountains without inconvenience, and, when he had scooped up some of the inhabitants on his thumb nail, was obliged to use a powerful microscope in order to see them. Voltaire should rather have gone to some of the most minute of the asteroids for his giant, for under the tremendous gravitation of such a world as he has described Micromegas himself would have been a fit subject for microscopic examination. But, however much we may doubt the stature of Voltaire's visitor from Sirius, we cannot doubt the soundness of the conclusion at which he arrived, after having, by an ingenious arrangement, succeeded in holding a conversation with some earthly philosophers under his microscope, namely, that these infinitely little creatures possessed a pride that was almost infinitely great.

East and south of Canis Major, which, by the way, is said to represent one of Orion's hunting dogs, is part of the old constellation Argo, which stands for the ship in which Jason sailed in search of the golden fleece. This giant constellation has since been split into three smaller star groups Carina, Puppis, and Vela. The observer will find many objects of interest in this region, although some of them are so close to the horizon in the northern latitudes that much of their brilliancy is lost. Note the two stars Zeta (ζ), called Naos, and Pi (π) in Puppis, near the lower edge of the map, then sweep slowly over the space lying between them. About half-way your attention will be arrested by a remarkable stellar arrangement, in which a beautiful half-circle of small stars curving above a larger star, which is reddish in color, is conspicuous. This neighborhood will be found rich in stars that the naked eye cannot see, including the open star cluster NGC 2451.

Just below the star Aludra, in Canis Major, is another fine group. In Puppis, the star Pi (π), which is deep yellow or orange, has three little stars above it, two of which form a pretty pair. The star Xi (ξ) in Puppis has a companion, which forms a fine test for binoculars, and is well worth looking for. Look also at the open cluster M93, just above and to the west of ξ. The stars Mu (μ). and Kappa (κ) are seen double with binoculars.

The two neighboring star clusters, M46 and M47, are very interesting objects in binoculars. A "fiery fifth-magnitude star," can be seen in the field at the same time. The presence of the Milky Way is manifest by the sprinkling of stars all about this region. In fact, the attentive observer will before this have noticed that the majority of the most brilliant constellations lie either in the Milky Way or along its borders. Cassiopeia, as we saw, sits athwart the galaxy whose silvery current winds in and out among the stars of her "chair"; Perseus is aglow with its sheen as it wraps him about like a mantle of stars; Taurus has the tips of his horns dipped in the great stream; it flows between the shining feet of Gemini and the head and shoulders of Orion as between starry banks ; the peerless Sirius hangs like a gem pendent from the celestial girdle. In the southern hemisphere we should find the beautiful constellation of Carina, containing Canopus, sailing along the Milky Way, blown by the breath of old romance on an endless voyage; the Southern Cross glitters in the thick star clouds of the southern sky; and the bright stars of Centaurus might be likened to the heads of golden nails pinning this wondrous scarf, woven of the beams of millions of tiny stars, against the dome of the sky. Passing back into the northern hemisphere we find Scorpio, Sagittarius toward the center of our galaxy, Aquila, the Dolphin, Cygnus, and resplendent Lyra, all strung along the course of the Milky Way.

Turning now to the constellation Monoceros (the Unicorn) we shall find a few objects worthy of attention. This constellation is of comparatively modern origin, having been formed by the son-in-law of the great Johannes Kepler. The region around the stars 8, 13, and 17 will be found particularly rich, and in dark sky you may see some nebulosity. Look also at the cluster M50, and compare its appearance with that of the clusters in Puppis.

With these constellations we finish our review of the stellar wonders that lie within the reach of so humble and inexpensive an instrument as binoculars. We have made the circuit of the sky, and the stars that illuminate the spring heavens are now seen advancing from the east, and pressing close upon the brighter stars of winter. Their familiar figures resemble the faces of old friends whom we are glad to welcome. These starry acquaintances never grow wearisome. Their interest for us is as fathomless as the deeps of space in which they shine. As we watch them in their courses, the true music of the spheres comes to our listening ears, the chorus of creation faint with distance, for it is by slow approaches that man draws near to it chanting the grandest of epics, the Poem of the Universe; and the theme that runs through it all is the reign of physical law.

Do not be afraid to become a star-gazer. The human mind can find no higher exercise. He who studies the stars will discover

*"An endless fountain of immortal drink
Pouring unto us from heaven's brink."*

The Moon

The Phases of the Moon

The Moon is our nearest neighbor in space — bright, barren, and bleak, and just a quarter million miles away, roughly 100 times closer than our next nearest neighbor, Venus.

While small in size—just ¼ the diameter of the Earth, the Moon's proximity makes it appear large, about 0.5 degrees across when full, and brighter than any other sight in the sky except for the Sun. It has hundreds of interesting features to see in a telescope. And even a modest pair of binoculars reveals dozens of craters, mountains, and smooth seas, or *maria*.

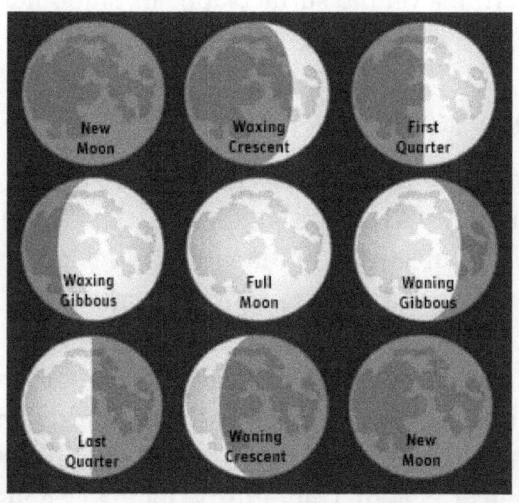

The phases of the Moon

The Moon revolves around the Earth each month, and as it does so, we see it go through *phases* (see Figure 1). When the Moon lies between the Earth and the Sun, it's in the "new Moon" phase and is not visible to us. Over the next few days, the Moon grows (or "waxes") to a crescent, then to first quarter, when it appears half-lit, roughly one week after the new Moon. Then its face becomes ever more lit by the sun and enters the gibbous phase until about two weeks after the new Moon, when it lies directly opposite the Sun in the sky and appears fully lit. Then the Moon's phases move in reverse, when they "wane" back through gibbous, last-quarter, and crescent phases, and back to new Moon again some four weeks after it was last new.   During the waxing phases, the Moon is visible mostly in the evening. As it wanes, it's best seen in daylight hours.

When you see the Moon partially illuminated by the Sun, the line between bright and dark is called the "terminator". Along this line, the craters, mountains and valleys stand out in stark relief caused by the long shadows cast by the Sun at lunar sunset and sunrise. Away from the terminator, the surface appears smoother, because the Sun casts shallower shadows.

The Landscape of the Moon

The Moon's largest and most visible features are its large, flat, gray patches called maria (MAH-ree-a), the Latin plural of mare (MAH-ray), or "sea." You can see the maria with the unaided eye. To some, these dark patches appear to form the outline of a face, or "man in the moon".

Of course, these maria are really not really seas or oceans. But early telescopic observers did not know this. The Italian astronomer Giambattista Riccioli gave them fanciful names such as Mare Tranquillitatis ("Sea of Tranquillity") and Oceanus Procellarum ("Ocean of Storms"), generally for the imagined astrological influences of the Moon's phases on the weather. Although astronomers soon realized the Moon has no water, the names of these surface features remained. Observation with telescopes and space probes revealed the "seas" are ancient lava flows that flooded most of the Moon's lowlands between 3.0 and 3.8 billion years ago.

The major maria are shown on Figure 2. Even the smallest binoculars will clearly show them. Learn their positions a little at time each night, and you will be on your way to understanding the geography of another world.

Because the maria are so smooth, they were selected as landing site for the manned Apollo lunar missions of 1969-1972. Figure 3 shows the position of each landing of Apollo 11 through 17; Apollo 13 did not land because of damage to its service module on the way to the Moon. You will not, of course, see any evidence of spacecraft landing or remnants through binoculars or even with the most powerful backyard telescope.

The dark "seas", or maria, *on the Moon's surface*

The side of the Moon seen in Maps 1 and 2 are the only sides we ever see from Earth. That's because the Moon rotates about its axis in the same period it takes to go around the Earth. This situation arose because of tidal effects by the Earth and Moon on each other. This "tidal locking" is seen on many moons throughout

Landing sites of the six manned Apollo missions

the solar system
The "far side" of the Moon was first imaged in the late 1950's by satellites. As Figure 4 shows, this side of the Moon appears quite different, with few maria and many more craters. The far side should not be confused with the dark side of the Moon; the far side essentially get just as much sunlight as the side of the Moon facing Earth.

Craters

The Moon is covered with thousands of craters, scars left by the impact of asteroids and comets over the past 4 billion years. While the Moon is too far away for us to see any craters with our unaided eye, even a small pair of binoculars reveals dozens of craters worthy of examination. A telescope reveals hundreds, if not thousands, more.

Most craters were formed during an era of "heavy bombardment" about 3.9 billion years ago by comets and asteroids left over from the formation of the solar system. Because the Moon has no air or water to erode the surface features, we see the craters fairly unchanged from the early days of the Moon. The Earth, too, was peppered with craters at the same time as the Moon, but those craters have eroded away over the eons. The moons maria have relatively few craters, which means these dark lava plains formed after the era of heavy bombardment.

The far side of the Moon

The large bright areas of the Moon— the lunar highlands — are the Moon's oldest terrain and are crusted with craters of every size, from dozens of miles wide down to tiny craterlets as small as a couple of hundred feet or less. In a telescope, you can even see craters within craters, or craters that overlap because a meteoroid smashed into an existing crater.

A telescope reveals that some large craters have a peak at their center caused by a rebound of the Moon's surface after a large impact. Other large craters look less like holes and more like walled plains, with dark, flat bottoms that flooded with lava after impact.

Young craters are surrounded by bright rays that extend radially across the surrounding landscape. Rays are great splashes of molten rock ejected by the impacts. Unlike craters and mountains, rays are best seen when illuminated more directly by the sun, away from the terminator. In binoculars, at full Moon, bright rays are visible extending from the large, young crater Tycho, which is only about 110 million years old.

The Moon also has a number of mountain ranges and individual peaks. Canyon-like cracks, or rilles, are sometimes visible, especially around the edges of maria.

Touring The Moon's Seas

Now that you know a little about the Moon's surface, let's embark on a descriptive tour of the major features visible through binoculars of magnification of 7-10x. At a magnification of 10x, the Moon appears just 25,000 miles away, similar to what an Apollo astronaut saw just hours before entering lunar orbit.

We will tour the moon's "seas" from east to west on the Moon, in order of appearance as the Moon waxes from new to full.
The apparently oval form of the Mare Crisium (Sea of Crises) (#8) is the first of the "seas" to come into sight after new moon. It makes a very striking sight for viewing at the waxing crescent

phase. It measures about two hundred and eighty by three hundred and fifty-five miles in extent and is surrounded by mountains, which can be readily seen when the sun strikes a few days after new moon. A spacecraft, Luna 24, landed in this region in 1976 and returned a soil sample to Earth.

West of Crisium, you can see the triangular region Palus Somnii (Marsh of Sleep) (#18); south of this region lies Mare Fecunditatis (the "Sea of Fertility or Fecundity") (#9). Unlike the smooth-edged Crisum, this sea is remarkable for its irregularly-shaped surface edges and the long crooked "bays" into which its southern extremity is divided.

The broad, dark-gray expanse of the Mare Tranquillitatis (Sea of Tranquility) (#7) is easily recognized. It looks mottled in some regions as a result of ridges and elevations. Apollo 11 landed in the extreme southwest of this sea on July 20, 1969, when astronauts Neil Armstrong and Buzz Aldrin became the first humans to walk on the Moon.

Mare Nectaris (Sea of Nectar) (#10) is connected with the Sea of Tranquility by a broad strait, while between it and the Sea of Fertility runs the lunar Pyrenees Mountains (#29), which are some twelve thousand feet high.

Mare Serenitatis (Sea of Serenity) (#6) lies northeast of the Sea of Tranquility. It's about four hundred and twenty miles across by four hundred and thirty miles long, being very nearly of the same area as the Caspian Sea on Earth. Serenity is deeper than the Sea of Tranquility, and is deepest toward the middle. Three-quarters of its "shore-line" are bordered by high mountains, and many isolated elevations and peaks are scattered over its surface. The last manned mission to the Moon, Apollo 17, landed at the eastern end of this sea. The Sea of Serenity is divided nearly through the center by a narrow, bright streak, apparently starting from the crater Tycho far in the south. This curious streak can be readily detected in binoculars.

Along the southern shore of the Sea of Serenity extends the high range of the Haemus Mountains (#27). South and southeast are the Mare Vaporum (Sea of Vapors) (#5), Sinus Medii (Bay of the Center) (#4), and Sinus Aestuum (Bay of Seething) (#3). At full moon, you may see three or four dark spots in the region occupied by these flat expanses. On the north and northwest of the Sea of Serenity is Lacus Somniorum (Lake of Sleep) (#17).

Mare Imbrium (Sea of Rains) (#2) is a fascinating region, not only in itself, but on account of its surroundings. Its level is very much broken by low, winding ridges, and it's variegated by numerous light-colored streaks. On its northeast border is the Sinus Iridum (Bay of Rainbows) (#16), upon which moon gazers have exhausted the adjectives of admiration. The bay is semicircular in form, one hundred and thirty-five miles long and eighty-four miles broad. Its surface is dark and level. At either end a splendid cape extends into the Sea of Rain, the eastern one being called Cape Heraclides, and the western Cape Laplace. They are both crowned by high peaks. Along the whole shore of the bay runs a chain of gigantic mountains, forming the southern border of a wild and lofty plateau, called the Sinus Iridum Highlands. Of course, a telescope is required to see the details of this magnificent lunar landscape, and yet much can be seen with binoculars. You may glimpse the capes at the ends of the bay projecting boldly into the dark, level expanse surrounding them, and the highlights of the bordering mountains sharply contrast with the dusky semicircle below. Two or three days after first quarter, the shadows of the peaks about the Bay of Rainbows may be seen. Sinus Roris (Bay of Dew) (#15) above the Bay of Rainbows, and the Mare Frigoris (Sea of Cold) (#1), are the northernmost of the dark seas visible from Earth.

Extending along the eastern side of the disk is the great Oceanus Procellarum (Ocean of Storms) (#14) southeast of which lie Mare Nubium (Sea of Clouds) (#11), Mare Humorum (Sea of Moisture)

(#12) and Mare Cognitum (Known Sea) (#13). These regions are irregular in outline, and broken by ridges and mountains. To the naked eye, even the relatively small Sea of Moisture is easily seen as a dark oval patch.

Major Craters and Mountains

Now, to lunar mountains and craters. The dark oval called Grimaldi (#70) can be detected on the western limb of the Moon by the unaided eye, although it requires a sharp vision and perhaps filter to reduce the glare of the moon. Grimaldi is simply a plain, a basin with worn down walls containing some fourteen thousand square miles. It's remarkable for its dark color. Grimaldi is also know to exhibit "transient lunar phenonmena" such as flashes of light and patches of color and haze. The reason for such activity is unknown and is an area of active research among dedicated amateur astronomers.

The basin called Schickhard (#62) is similar to Grimaldi, nearly as large, but it does not possess the same dark tint in the interior. The huge mountains around Schickhard make it a fine sight when the sun is rising upon them shortly before full moon.

Other than the maria, the feature of the full Moon's surface that instantly attracts attention is the remarkable brightness of the southern part of the disk, and the brilliant streaks radiating from a bright point near the lower edge. Of this region, many observers remark the Moon looks like a peeled orange. The bright point, which is the great crater Tycho, looks exactly like the pip of the orange, and the light streaks radiating from it in all directions bear an striking resemblance to the streaks that one sees upon an orange after the outer rind has been removed.

Tycho (#60) is the most famous of the Moon's crater, though it's not the largest. It's about fifty-five miles across and three miles deep. In its center is a peak five or six thousand feet high. Tycho is the radial point of the great rays that are caused by ejected molten rock thrown out by the impact of a comet or asteroid. Unlike many details on the Moon's surface, the rays are best seen at full Moon; they cannot be seen at all until the sun has risen to a certain elevation above them. The rays pass straight over the most rugged regions of the moon for some 2,000 kilometers, retaining their brilliance as far away as the Sea of Serenity. Some of the rays can be seen on parts of the Moon illuminated only by the reflected light from the Earth.

The double chain of great crater-plains reaching halfway across the center of the moon contains some of the grandest configurations of mountain, plain, and crater. The names of the principal ones can be learned from the map, and you will find it interesting to watch them come into sight about first quarter, and pass out of sight about third quarter. With binoculars, some of them look like enormous round holes in the inner edge of the illuminated half of the moon. Theophilus (#55), and nearby Cyrillus and Catharina are three of the finest walled plains on the Moon.

The Caucasus Mountains (#25) are a mass of highlands and peaks which introduce you to a series of formations resembling those of the mountainous regions of the earth. The highest peak in this range is about nineteen thousand feet high. Between the Caucasus and the Apennines (#26) lies a level pass, or strait, connecting the Sea of Serenity with the Sea of Rains. The Apennines are the greatest of the lunar mountains. They extend some four hundred and sixty miles in length, and contain one peak twenty-one thousand feet high; many others vary from twelve thousand to nearly twenty thousand feet, much higher than the Apennines of the earth. As this range runs at a considerable angle to the line of sunrise, its high peaks can be seen tipped with sunlight for a long distance beyond the generally illuminated edge about the time of first quarter. Even with the naked eye the sun-touched summits of the lunar Apennines may at that time be detected as a tongue of light projecting into the dark side of the moon. The Alps (#23) are another mountain range of great elevation, whose highest peak is a good match for the Mont Blanc of Earth, after which it has been named.

Plato (#80) is a celebrated dark and level plain, surrounded by a mountain ring, and presenting in its interior occasional changeable phenomena which have caused much speculation, but which, of course, lie far beyond the reach of binoculars.

The great crater Copernicus (#74) bears a general resemblance to Tycho, and is slightly greater in diameter but is not quite so deep. It has a cluster of peaks in the center, whose tops may be seen with binoculars as a speck of light when the rays of the morning sun, slanting across the valley, illuminate them while the floor of the crater lies in darkness. Copernicus is the center of a system of rays that resemble those of Tycho, but are very much shorter.

Andrés Valencia - Observatorio ARVAL http://www.arval.org.ve

North:
- 1- Mare Frigoris (Sea of Cold)
- 2- Mare Imbrium (Sea of Rains)
- 3- Sinus Aestuum (Bay of Seething)

Northeast:
- 4- Sinus Medii (Bay of the Center)
- 5- Mare Vaporum (Sea of Vapors)
- 6- Mare Serenitatis (Sea of Serenity)

7- Mare Tranquillitatis (Sea of Tranquillity)

8- Mare Crisium (Sea of Crises)

17- Lacus Somniorum (Lake of Sleep)

18- Palus Somnii (Marsh of Sleep)

19- Mare Anguis (Sea of Snakes)

20- Mare Undarum (Sea of Waves)

Southeast:

9- Mare Fecunditatis (Sea of Fecundity)

10- Mare Nectaris (Sea of Nectar)

21- Mare Spumans (Sea of Foam)

Southwest:

11- Mare Nubium (Sea of Clouds)

12- Mare Humorum (Sea of Moisture)

13- Mare Cognitum (Known Sea)

22- Palus Epidemiarum (Marsh of Diseases)

West:

14- Oceanus Procellarum (Ocean of Storms)

Northwest:

15- Sinus Roris (Bay of Dew)

16- Sinus Iridum (Bay of Rainbows)

Mountains

Northeast:

23- Montes Alpes

24- Vallis Alpes (Alpine Valley)

25- Montes Caucasus

26- Montes Apenninus

27- Montes Haemus

28- Montes Taurus

Southeast:

29- Montes Pyrenaeus

Southwest:

'30- Rupes Recta (Straight Wall) [Geological Fault]

31- Montes Riphaeus

Northwest:

32- Vallis Schröteri (Schröter's Valley) [Northwest of Crater Aristarchus, 73, and North of Crater Herodotus]

33- Montes Jura

Craters

Northeast:

34- Crater Aristotle [on the East part of Mare Frigoris, 1]

35- Crater Cassini

36- Crater Eudoxus

37- Crater Endymion

38- Crater Hercules

39- Crater Atlas

40- Crater Mercurius

41- Crater Posidonius

42- Crater Zeno

43- Crater Le Monnier

44- Crater Plinius

45- Crater Vitruvius

46- Cráter Cleomedes

47- Crater Taruntius

48- Crater Manilius

49- Crater Archimedes

50- Crater Autolycus

51- Crater Aristillus

Southeast:

52- Crater Langrenus

53- Crater Goclenius

54- Crater Hypatia

55- Crater Theophilus

56- Crater Rhaeticus [Crater Hipparchus is directly South of Crater Rhaeticus]

57- Crater Stevinus
58- Crater Ptolemaeus
59- Crater Walter

Southwest:
60- Crater Tycho
61- Crater Pitatus
62- Crater Schickard
63- Crater Campanus
64- Crater Bulliadus
65- Crater Fra Mauro
66- Crater Gassendi
67- Crater Byrgius
68- Crater Billy [Mons Hansteen is to the North of Crater Billy]
69- Crater Crüger
70- Crater Grimaldi
71- Crater Riccioli

Northwest:
72- Crater Kepler
73- Crater Aristarchus [Crater Herodotus is West of Crater Aristarchus]
74- Crater Copernicus
75- Crater Pytheas
76- Crater Eratosthenes [near the Southwestern extreme of Montes Apenninus, 26]
77- Crater Mairan
78- Crater Timocharis
79- Crater Harpalus [Crater Pythagoras is North of Crater Harpalus]
80- Crater Plato

The Planets

When attempting to view the planets with binoculars, don't expect too much. The features of the surfaces and cloud-tops of the planets are far beyond the capabilities of even powerful binoculars. But the difference between the appearance of a large planet and that of the stars will certainly be obvious.

Unlike the stars, planets move about in the sky from week to week and month to month as they and the Earth revolve around the sun. In fact, planets take their name from the ancient Greek term, "wanderer". To find where each planet lies in the sky, you can check online resources such as Sky and Telescope, or printed yearly guides of the position of the Moon and planets, such as that published by the Royal Astronomical Society of Canada.

Mercury

Mercury, the closest planet to the sun, rapidly changes its place in the sky from week to week, even from day to day. Not many people ever see Mercury; even the great Nicholas Copernicus was said to never have seen the planet in his lifetime. A powerful backyard telescope shows no surface detail on Mercury... it's too far away and lies to close to the sun. But compared to viewing with the unaided eye, the beauty of the planet is greatly increased when viewed with binoculars. Mercury is brilliant enough to be readily distinguishable, even at twilight. When it's far enough from the sun in the sky for viewing, you may have fine views of this tiny baked world, glittering like a globule of shining metal through the fading curtain of a sunset or the onset of a sunrise.

Venus

Venus is, under favorable circumstances, much more interesting for binocular observations. Because they lie closer to the sun than the Earth, Venus, like Mercury, goes through phases much like the Moon. The crescent phase, where the angular span of the planet is largest, can be glimpsed with a powerful glass when the planet is nearest the Earth and sun (a position called inferior conjunction). Even when the form of the planet cannot be seen, its brilliance makes it an attractive sight. Venus is embedded in a dense blanket of permanent cloud cover that adds to the reflectivity of the planet, but which means you'll see no surface features. Just watching the dance of Venus across the sky as it moves cyclically from evening "star" to morning "star" and back again is a delight in itself. Aside from the sun and the Moon, Venus is the brightest object in the night sky.

Jupiter and its four largest moons

Mars

Mars is smaller and more distant than Venus, and it's too far away to display surface detail in binoculars. Yet when it's at or near opposition, that is, opposite the sun in our sky and closest to Earth, it is a superb object even for a binoculars because of the contrast its deep reddish-yellow color presents with most stars. It can often be seen close to the moon and stars, and the beauty of this proximity is in some cases greatly enhanced in binoculars.

Jupiter

Jupiter, although much more distant than Mars, is a far more conspicuous object in the sky because of its immense size. Binoculars offer too little magnification to show the banded clouds and features such as the famed "Red Spot", which is an immense hurricane-like storm that has persisted for hundreds of years. But even a modest glass will show Jupiter's four moons, which, as they revolve around the giant planet, present a miniature of the solar system.

If you can't spot all four of Jupiter's largest moons, chances are one or more of them lies either between you and the planet, or behind the planet, or buried in the planet's shadow, or else so close to the planet as to be concealed by its brilliance. The positions of the moons, named Io, Europa, Ganymede, and Callisto, can be found in the print or online edition of Sky and Telescope and other almanacs.

Of course, since the motions of Jupiter's satellites, particularly of the inner ones, are very rapid, their positions are continually changing, and their configurations are different each night, and indeed, from hour to hour. If you have any doubt about which moon is which, or think they may be little stars, you have only to carefully note their position and then look at them again the next evening. You may even notice their motion in the course of a single evening, if you begin early and follow them for three or four hours. It is impossible to describe the peculiar attractions of the scene presented by the great planet and his four little moons on a serene evening to an observer armed with good binoculars. Probably much of the impressiveness of the spectacle comes from the knowledge that those little points of light, shining in a row are, at every instant, under the gravitational influence of Jupiter, and that as we look upon them, obediently making their revolutions, never venturing beyond a certain distance away, we behold the forces to which our own planet is subject as it revolves around the sun, to whose control even Jupiter in his turn submits.

Saturn

The beautiful planet Saturn requires for the observation of its rings magnifying powers far beyond those of binoculars. It would be well, however, for you to trace its slow motion among the stars with the aid of an almanac. You may be able to see, under favorable circumstances, the largest of its eight moons, Titan, which has a brightness of magnitude 8.5: faint, but still within the range of most binoculars.

The Outer Planets

It may appear somewhat presumptuous to place Uranus, a planet which required the telescope and the eye of William Herschel to discover, in a list of objects for binoculars. But Uranus was likely seen many times before Herschel's discovery, being simply mistaken, on account of the slowness of its motion, for a fixed star. When near opposition from the sun, Uranus looks as bright as a sixth-magnitude star, and can be easily detected with the naked eye when its position is known. With binoculars, this distant planet can be watched as it moves deliberately onward in its gigantic orbit. Its passage by neighboring stars is an exceedingly interesting sight; it's in this way that you may most easily find the planet. Armed with binoculars, a star map, and positional information from an almanac or periodical such as Sky and Telescope, you can discern the location of Uranus and assure yourself of the existence of the planet by watching its motion against the background stars from night to night.

Neptune, alas, is beyond the reach of all but the most diligent binocular observer. It can be found in the same way as Uranus, but at magnitude 8 it shines more than 6x fainter than Uranus, so a careful eye is required to spot this distant world at all.

Pluto, whether you consider it a planet or not, if far beyond the reach of binocular observers, and indeed most backyard telescopes.

Appendix 1 - A Word About Star Names

Bright stars such as Rigel, Sirius, Aldebaran, and Capella were named by classical Greek and Arabic astronomers thousands of years ago. But fainter stars went unnamed until Renaissance astronomers and their heirs began to formalize star names in each constellation by assigning lower-case Greek letters in order of brightness. So, for example, the brightest star in Cygnus, Deneb, is assigned the name α Cygni (alpha Cygni), and the second bright star in Cygnus, Albireo, is assigned the name β Cygni (beta Cygni), and so on, through gamma and delta and the entire Greek alphabet. If you're rusty with Greek, a summary of the letters is found on the next page.

In some cases, either because of error or otherwise, the stars do not always follow the order of brightness. Castor is the brightest star in Gemini, for example, yet it's labeled β (Beta) Geminorum. It's a little confusing at first, but the designations have stuck.

As the sky was mapped more fully, astronomers ran out of Greek letters and turned to numbers for the stars, then finally to a more complex scheme based on formal catalogs of the stars made by various astronomers. But that's more than you need to know. Just remember than each star has a name and number, and its position is faithfully reproduced for your benefit on star maps.

α	alpha	ν	nu
β	beta	ξ	xi ('shi')
γ	gamma	o	omicron
δ	delta	π	pi

ε	epsilon	ρ	rho
ζ	zeta	σ	sigma
η	eta	τ	tau
θ	theta	υ	upsilon
ι	iota	φ	phi
κ	kappa	χ	chi ('ki')
λ	lambda	ψ	psi ('si)
μ	mu	ω	omega

Appendix 2 - Stellar Magnitudes

To describe the brightness of objects in the sky, astronomers often use a numerical measure called "magnitude".

In this system, first worked out by ancient Greek astronomers, brighter stars and planets have a smaller numerical value of magnitude than fainter objects. So, for example, a star with magnitude 4 is brighter than a star with magnitude 5.

To be more exact, an object with magnitude 1.0 is 100 times brighter than an object with magnitude 6.0. So each step of 1.0 in magnitude is the fifth root of 100. That means a star of magnitude 3.0 is 2.512 times as bright as a star of magnitude 4.0, which is 2.512 times as bright as a star of magnitude 5.0, and so on. Try it yourself, if you have a calculator handy.

With your naked eye, you can see objects down to 6th magnitude; with a pair of 7×50 binoculars you can see down to 10.5 or so; and with an 8-inch telescope, perhaps 13.5. Using sophisticated cameras and software, the Hubble can detect objects to about 30th magnitude... about 4 billion times fainter than you can see with your eye.

An object brighter than 0th magnitude has a negative magnitude; the brightest star, Sirius has an apparent magnitude -1.4; the full moon has apparent magnitude -13, and the Sun has apparent magnitude of -26.

Usually, backyard stargazers talk about "apparent" magnitude, which measures how bright a star appears in the sky, regardless of how bright is truly is.

But "absolute" magnitude is a measure of the true, intrinsic brightness of a star. It's defined as the apparent magnitude of an object if it was 32.616 light-years away

So while the sun has an apparent magnitude of -26, if we could see it at a distance of 32.616 light-years, it would shine at a very modest magnitude 4.7.

Deneb, the brightest star in Cygnus, has an absolute magnitude of -6.93, some 40,000 times as bright as our Sun. But its apparent magnitude is only 1.25 because it's so far away, roughly 3,200 light-years from Earth.

Here are some apparent and absolute magnitudes of the 25 brightest stars as seen from Earth, other than the Sun, for reference:

Star Name	Designation	Apparent Magnitude	Distance (light years)	Absolute Magnitude
Sirius	Alpha CMajoris	-1.44	8.6	1.45
Canopus	Alpha Carinae	-0.62	309	-5.53
Rigel Kent	Alpha Centauri	-0.28	4.32	4.11
Arcturus	Alpha Bootes	-0.05	36.7	-0.35
Vega	Alpha Lyrae	0.03	25	0.6
Capella	Alpha Aurigae	0.08	4208	-0.51
Rigel	Beta Orionis	0.18	863	-6.93
Procyon	Alpha CMinoris	0.4	11.5	2.67
Achernar	Alpha Eridini	0.45	139	-2.7
Betelguese	Alpha Orionis	0.45	498	-5.47

Star Name	Designation	Apparent Magnitude	Distance (light years)	Absolute Magnitude
Hadar	Beta Centauri	0.61	392	-4.79
Altair	Alpha Aquilae	0.76	16.7	2.21
Acrux	Alpha Crucis	0.77	322	-4.2
Aldebaran	Alpha Tauri	0.87	66.6	-0.68
Spica	Alpha Virginis	0.98	250	-3.44
Antares	Alpha Scorpii	1.05	554	-5.09
Pollux	Beta Geminorum	1.16	33.8	1.08
Fomalhaut	Alpha P. Austr.	1.16	25.1	1.74
Mimosa	Beta Crucis	1.25	279	-3.41
Deneb	Alpha Cygni	1.25	1412	-6.93
Regulus	Alpha Leonis	1.36	79.3	-0.57
Ahdara	Epsilon CMajoris	1.5	405	-3.97
Castor	Alpha Geminorum	1.58	50.9	0.61
Gacrux	Gamma Crucis	1.59	88.6	-0.58
Shaula	Lambda Scorpii	1.62	571	-4.6